Proceedings of the 30th International Geological Congress
Volume 20

Geophysics

Proceedings of the 30th International Geological Congress

PROCEEDINGS OF THE
30TH INTERNATIONAL GEOLOGICAL CONGRESS

BEIJING, CHINA, 4 - 14 AUGUST 1996

VOLUME 20

GEOPHYSICS

EDITOR:
LIU GUANGDING
INSTITUTE OF GEOPHYSICS, CHINESE ACADEMY OF SCIENCES, BEIJING, CHINA

CRC Press
Taylor & Francis Group
Boca Raton London New York

CRC Press is an imprint of the
Taylor & Francis Group, an **informa** business

First published 1997 by VSP BV Publishing

Published 2017 by
CRC Press
Taylor & Francis Group
6000 Broken Sound Parkway NW, Suite 300
Boca Raton, FL 33487-2742

© 1997 by Taylor & Francis Group, LLC
CRC Press is an imprint of Taylor & Francis Group, an Informa business

First issued in paperback 2019

No claim to original U.S. Government works

ISBN 13: 978-0-367-44827-1 (pbk)
ISBN 13: 978-90-6764-241-5 (hbk)

Visit the Taylor & Francis Web site at
http://www.taylorandfrancis.com

and the CRC Press Web site at
http://www.crcpress.com

CONTENTS

Proc. 30th Int'l Geol. Congr., Vol.20, pp. 1-10
Liu (Ed.)
© VSP 1997

Basement Tectonics in Continental China, Based on Regional Geophysical Interpretation

WANG MAO-JI

Aero Geophysical Survey and Remote Sensing Center, MGMR, China

Abstract

Regional geophysical studies made a fundamental contribution to the recognition of basement tectonics in continental China. The gravity trend patterns and the magnetic features are used to map the Precambrian basement. The extensive paired gravity anomalies, separating terranes with contrasting structural character-istics, afford a method for identifying the location of a suture and the associated subduction direction. Geophysical transects across several suture zones support this evidence. The regional isostatic gravity anomalies and free air gravity anomalies provide insight into the large scale crustal structure and the con-temporary plate dynamics.

Key words: Reginonal geophysics, Basement tectonics, Continental China

INTRODUCTION.

Current knowledge of the basement tectonics of China is guite diverse. The newly compiled gravity and aeromagnetic maps have made significant advances in understanding the basement tectonics. Gravity and magnetic data, processed through appropriate filtering and analysis, are integrated with geological and other geophysical data to define the major basement structures. The goals of present studies have been to map the Precambrian basement, to recognize the paleo—suture zones, and to determine the nature of ancient tectonic regime. The new concepts of the basement tectonic framework and major tectonic features have also been proposed.

GRAVITY AND AEROMAGNETIC DATA SOURCES

The 1:50,00,000—scale gravity and aeromagnetic maps of China were compiled using data bases then available. Aeromagnetic data are available from Aero Geophysical Survey(AGS), Ministry of Geology and Mineral Resources(MGMR) with line spacing varying from 500, 1000—2000m to as large as 5—20 km, and with ground clearances from 100—200m to 300—600m, and to more than 1000m in some mountain areas [1]. For the purpose of digital analysis, the data were inter-polated from original survey data set onto a 5×5km grid. In the process, the magnetic level for the individual areas was adjusted, and the IGRF was removed from the data. Gravity data from various sources were compiled by Methodological and Technical Center of Regional Gravity

Survey, MGMR. Collected data obtained from surveys on different scales involves about 300,000 measurements. About 20,000 gravity observations have been selectecd with data coverage of about 378km^2 per station in the eastern part and 644km^2 per station in the western part of China[1]. The data set was interpolated to 10×10km grid serving as the basis for digital analysis. Both gravity and magnetic data were contoured by computer and were plotted as color relief images.

MAPPING OF PRECAMBRIAN BASEMENT

We use residual gravity anomalies and horizontal gravity gradients to define gravity trends and domains. The trends can be seen to reflect the strike and direction of the major lithologic units and the domains define the major regions of coherent structure within the crust. The anomaly trends are of most use in defining basement structure, where they are of large amplitude and have a well–defined pattern[2,3].

On the anomaly trend map (Fig.1), the relative ages of adjacent blocks can be determined and

Figure 1. Gravity trend map derived by tracing axes of local gravity anomalies. Boundaries of gravity domains are shown as thick lines where marked by major gravity gradients (paired gravity anomalies), and as a

---

[1] Bouguer gravity map of China on the scale of 1:2,500,000, Explanation notes, Institute of Geophysical and Geochemical Exploration, MGMR(1989).

the Precambrian basement are inferred. The block in which the trends are truncated by the boundary is inferred to be older than the boundary, and the block with trends parallel to the boundary is inferred to be younger than the other block, and the same age as the boundary. The Precambrian block is characterized essentially by the marked diversity of trends, such as Junggar block (XIV), Tarim block(XII), Alashan and Xining –Lanzhou block(VII), etc. On the basis of the magnetic features (amplitude, shape, and orientation), four types of Precambrian basement corresponding to different geologic ages have been recognized, namely magnetic(Ar–Pt$_1$), medium magnetic (Pt$_{1-2}$), weakly magnetic (Pt$_{1-3}$), and nonmagnetic (Pt$_3$) basement. Figure 2 shows the distribution of magnetic basement in China.

Figure 2. Map showing distribution of magnetic basement (Ar–Pt$_1$) in China. Hatched areas indicate the regions with magnetic basement: 1. Junggar–Hami, 2. Ili, 3. South Tarim, 4. Dunhung, 5. Qaidam, 6. Xining–Langzhou, 7. Alashan, 8. North Ordos, 9. Yanshan–Luliangshan, 10. He–Huai, 11. Qingling–Dabie, 12. Yangtze.

GEOPHYSICAL SIGNATURE OF PALEO–SUTURE ZONE

The extensive positive–negative paired gravity anomalies occur along a number of boundaries between major crustal blocks. The anomalies are of a higher amplitude than the adjacent anomalies, and many are very elongate. They are generally considered to be a signature of collisional suture zone[4–6]. Gravity gradients presently recognized to be paired anomalies overlying a crustal block boundary are shown in Figure 1. Studies have shown that the gravity high is generally on the younger terrane, the negative component always straddles the boundary or lies largely within the older crust, so the contact generally dips towards the gravity high[6]. When the younger crust thrusts against the older crust it will tend to be deformed, becoming thickened and denser. Although the polarity of the anomaly pair with respect to relative ages of sutured terranes is consistent, the difference in gravity signature may result from two differing kinds of collision: continent / continent and continent / arc collsion[4]. In some places, magnetic signature at structural boundaries is also evident where the belts of positive magnetic anomaly correlates with calc—alkaline magmatic arc.

The application of paired gravity and magnetic anomalies together with the geophysical transects to mapping sutures can be demonstrated with several examples:

Karamali suture zone (Fig.3) This suture is expressed by complex belt of positive and negative magnetic anomalies. The existence of ophiolite melange belt and the "I" type plutons[7] all along the positive anomalies supports the collision between the Junggar plate and the Siberian plate.

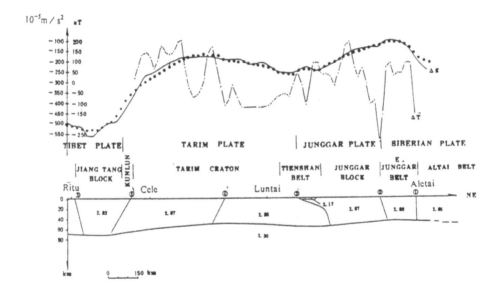

Figure 3. Simple density model for the transect from Aleitai to Ritu, across Siberian, Junggar, and Tarim plates. Density unit in T / m³. Solid circles denote the calculated gravity anomaly. Major tectonic elements. (1)Irtysh fault, (2)Karamali suture, (3)mid—Tianshan suture, (4)mid—Tarim fault, (5)West Kunlun suture, (6)Banggong—Nujiang suture.

Mid—Tianshan suture zone (Fig.3) This suture is located along the southern margin of mid—Tianshan mountain and expressed by paired gravity anomaly. In response to convergent interaction between the Tarim and the Junggar blocks, a magmatic arc developed[9] producing a belt of positive magnetic anomaly. The island arc and continental block collision system passed into a continent / continent collsion zone between the Tarim plate and the Kazakhstan—Junggar plate.

West kunlun suture zone (Fig.3) Along the southern part of west kunlun mountain, the position of the suture is expressed by a steep gravity gradient. There, gravity values are higher over the thinner and denser Tarim block. This suggests a collision between the Tarim block and the North Kunlun island arc.

Karakunlun–Lancangjiang suture zone (Fig.4) This sature zone locates on the northern edge of the Gondwanaland and is marked by paired gravity anomaly zone and by the changes of gravity trend pattern. The negative component is directly over the south–dipping subduction zone. It portrayed an active margin of the Tibet plate collided with the South China plate.

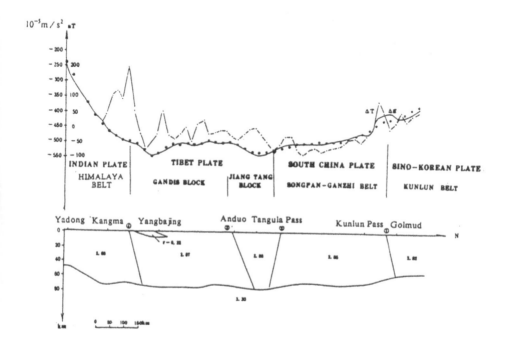

Figure 4. Simple density model for the transect from Yartong to Golmud, across North China, South China, and Tibet plates. Density unit in T/m³. Soli; circles denote the calculated gravity anomaly. Sutures: (1)Kunlun–Qingling, (2)Karakunlun–Lancangjian;, (3)Banggong–Nujiang, (4)Yarlung Zhangbo.

Yarlung Zhangbo suture zone (Fig.4) This suture zone is characterized by a steep gravity gradient and an intensive positive magnetic anomaly zone due to the effect of Gandis volanic arc and Yarlung Zhangbo ophiolite belt. The crust of the Proterozoic Tibet block is thicker and denser than that of the Archean Indian craton. These signatures marked a major suture zone resulted from the continent / continent collision between India and Tibet.

MEGA GRAVITY LINEAMENTS

A map showing isostatic gravity anomalies of wavelength from 500 to 1000km is given in Figure 5. Seven notable NW–trending mega lineaments characterized by alternation of positive and

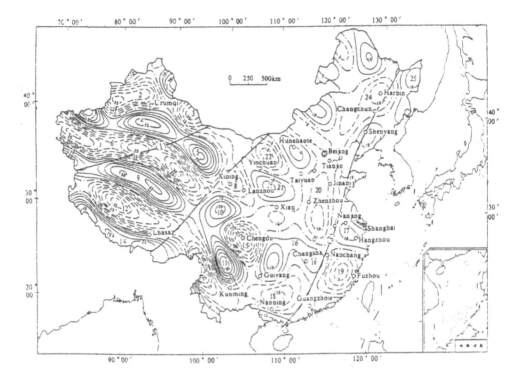

Figure 5. Medium—wavelength isostatic gravity map of China. Positive anomaly is marked by solid line, negative anomaly by dashed line. Contour interval is $5 \times 10^{-5} \mathrm{m/s^2}$. Thick lines denote the major shear zones. The location of major tectonic features: 1. Altai uplift, 2. Junggar—Ili basin, 3. Hami—Santanghu depression, 4. Tianshan mountain, 5. Qilian mountain, 6. South Tarim basin, 7. Qaidam basin, 8. Xining—Langzhou depression, 9. Kunlun mountain, 10. Animaging mountain, 11. Bayankala mountain, 12. Tibet block, 13. West Yunnan uplift, 14. Himalavas, 15. Sichuan basin, 16, 17. Jiangnan uplift, 18. Youjiang trough, 19. Zhejiang—Fujian volcanic area, 20. North China depression, 21. Ordos—Alashan block, 22. Huanghe graben, 23. Jiaoliao uplift, 24. Sungliao depression, 25. Jiamusi uplift. Major shear zones: (1) Altun, (2)Longshan—Linzhi, (3)Tanlu.

negative anomalies in the western part of China seem to be planiform. These are Altai, Junggar—Hami, S. Tarim—Qaidam, Kunlun—Bayankala, Tibet—W.Yunnan, and Himalaya lineaments. Theoretical stress studies suggest that these anomalies are far too great in lateral extent to be explained by masses supported by the lithosphere. It is therefore postulated that they are possible relics of the earth's attempt to restore isostatic equilibrium[9]. If so, they represent fundamental division of the lithosphere and indicate the nature of ancient tectontic regime. Their boundary is generally a suture zone.

Three regional lineaments across China's continent representing huge shear zones have also been identified in this map. These are the Tanlu fault in the east, the Longshan—Linzhi fault in the middle, and the Altun fault in the west. Matching of major anomalies across these lineaments suggests left lateral slip nature of these faults.

REGIONAL FREE AIR GRAVITY ANOMALIES AND THEIR RELATION TO TECTONICS

A proper and easily treated gravity quantity for tectonic analysis is the free air gravity anomaly as this is simply the observed gravity value corrected for elevation and latitude. Gravitational attraction due to all mass remains. The free air gravity anomalies of both local and regional scale yield important insight into the forces that are affecting contemporary plate action and paleointeraction. On the long-wavelength ($\lambda > 1000$km) free air gravity map (Fig.6) four prominent gravity anomalies are present. These are a negative anomaly in Northwest China, a positive anomaly in North and Northeast China, a negative anomaly in South China, and a positive anomaly over Qinhai–Tibet region. It is possible that they are caused by deep–seated mass concentration in the mantle, which may be related to mass distributions in an early earth, or by the contemporary stress and thermal regime. The broad positive anomaly is interpreted as being due to regional compression with subsequent regional uplift, the negative anomaly persisting over stable blocks is due to regional subsidence. These suggest that the Junggar, Tarim, Qaidam, Alashan, and Yangtze blocks within the negative anomaly zone had probably been an unified continental block in early Proterozoic time; the Songliao block in northeastern China could be genetically related to the North China craton. Apparently, it couldn't be considered that the traditional concept of the Tarim–North China proto continent is tenable.

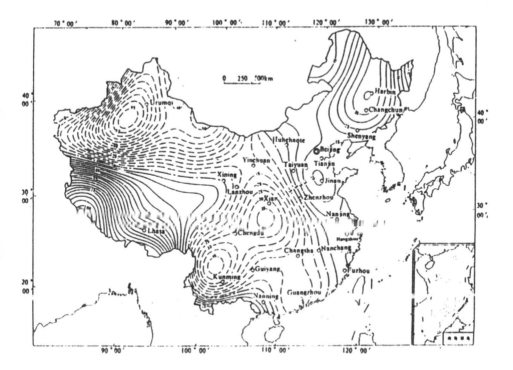

Figure 6. Long–wavelength free air gravity map of China. Positive anomaly is marked by solid line, negative anomaly by dashed line. Contour interval is 5×10^{-5} m / s^2.

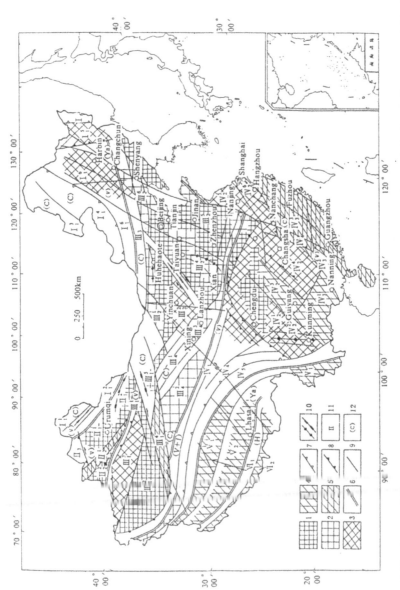

Figure 7. Sketch map of basement tectonics in China. Precambrian basement: (1) magnetic (Ar—Pt$_1$), (2) medium magnetic(Pt$_{1-2}$), (3)weakly magnetic(Pt$_{1-3}$) (=) nonmagnetic(Pt$_{1-3}$), (5) reworked block; (6) paleo—suture zone; (7) paleo—subduction zone; (8) thrust zone; (9) major fault; (10) fold belts, the letter in parenthses represents folding period.

BASEMENT TECTONIC FRAMEWORK OF CHINA

The sketch map of basement tectonics of China is shown in Figure 7. The construction of basement tectonic framework was accompanied by recognition of major tectonic elements, among these are the ancient continental margins, the distribution of Precambrian blocks of different ages, the major faults, and the other elements. Of special importance in this respect are the following proposals (Fig.8): 1) China's continent comprises a mosaic of several distinct continental blocks separated by accretionary fold belts, 2) the Tianshan and Qilianshan could be linked within an unified orogenic belt in early Paleozoic, 3) the Alashan block was rifted away from the Qaidam block during the opening of the Qilian sea trough, 4) the Xining—Langzhou block, intervening between Qilianshan and north Qingling mountain, was detached from the Qaidam block by the Longshan—Linzhi left—lateral slip fault, 5) the Tarim and Qaidam blocks were originally a single block, dissected by the Altun strike slip fault possibly in late Proterozoic, 6) the southeast China has been a paleo—continental block[10–11] rifted away from and subsequently juxtaposed with the Yangtze block along Shaoxing—Pinxiang—Chaling—Beihai boundary in late Proterozoic, and has been reworked in several episodes since Paleozoic and Mesozoic time. The crustal reworking can be manifested by complex structures with several cycles of extension and major intrusive and volcanic activity and by change in anomaly trends at the block boundary and in the rifts traversing blocks.

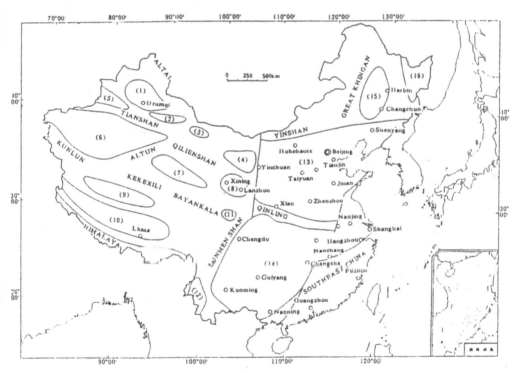

Figure 8. Sketch map showing possible Precambrian blocks and principal fold belts. The Precambrian blocks are numbered in paretheses: (1) Junggar, (2) Turfan, (3) Mazonshan, (4) Alashan, (5) Ili, (6) Tarim, (7) Qaidam, (8) Xining—Langzhu, (9) Jiangtang, (10) Gandis, (11) Zolge, (12) Baoshan, (13) North China, (14) Yangtze, (15) Sungliao, (16) Jiamosi.

CONCLUSIONS

In China's continent, the main features of basement blocks and their covers are the consequence of well defined episodes of basement reactivation which are reflected in the pattern of anomaly trends, in the paired gravity anomalies, and in aeromagnetic data. Two kinds of gravity anomalies are more useful in tectonic analysis. One is short—and long—wavelength free air gravity anomalies caused by ancient tectonic features and by deformation of lithosphere, another is isostatic gravity anomalies stronly related to the known geologic division.

According to the available data, we have proposed new recognition on the basement tectonic framework of China, among which are recognition of the paleosuture zones, of the Precambrian blocks and their tectonic attributions, and of the major tectonic features. The recognition of ancient tectonic regime is of special importance.

It has to be stressed that the basemeint tectonics of China presented here is very basic and general. As more detailed geophysical data become available, especially the high—resolution seismic reflection profiling are more widely utilised, the new constrains will be made available for re—evaluating gravity and aeromagnetic data and an understanding of continental tectonics should be further enhanced.

REFERENCES

1. Aero Geophysical Survey, MGMR. *Technical description for the* 1:4,000,000 *aeromagnetic anomaly map of China and adjacent sea areas*, Beijing: China Cartographic Publishing House (1989).
2. M.D. Thomas, R.A.F. Grieve, V.L.Sharpton. Structural fabric of the North American continent, as defined by gravity trends, *International conference on Basement Tectonics* V.1,257—276, Dordrect: Klacwer Academic Publisher(1987).
3. P. Wellman. Gravity trends and the growth of Australia: a tentative correlation, *J. Geol. Soc. Austr.* 23, 11—14(1976).
4. M.D. Thomas. Ancient collisional continental margins in the Canadian Shield: Geophysical signatures and derived crustal transects, *International Conference on Basement Tectonics*, V.2, 5—25, Dordrect: Kluwer Academic Publisher (1988).
5. M.D. Thomas. Tectonic significance of paired gravity anomalies in the southern and central Appalachians, *Geol. Soc. Amer. Memoir*158, 113—124 (1983).
6. P. Wellman. Development of the Australian Proterozic crust as inferred from gravity and magnetic anomalies, *Precambrian Res.* 40 / 41, 89—100 (1988).
7. Cai Wen—Jun. Preliminary study on plate tectonics of northeastern Junggar in Xinjiang, *Contributions to the plate tectonics of northern China* 1, 1—26, Beijing: Geological Publishing House (1986).
8. Hu Shou—Xi, Guo Ji—Chun, Gu Lian—Xi, et al. Geological features of Caledonian Orogenic belt and its importance in the tectonic framework of the E. Tianshan district, *Geoscience of Xinjiang* 1, 32—45 (1990).
9. K.F. Sprente, E.R.Kanasewich. Gravity modelling and isostasy in western Canada, *J. Can. Soe. Expc. Geophys.* 18, 49—57 (1982).
10. Zhang Wen—Yu. *Outline of tectonics of China*, Beijing: Science Press (1959).
11. Shi Tao, Xu Bu—Tai, Liang Ru—Hua, et al. *Metamorphic basement geology of Zhejiang and Fujian prorinces in China*, Beijing: Science Press (1988).

Proc. 30th Int'l Geol. Congr., Vol. 20, pp. 11-20
Liu (Ed.)
© VSP 1997

DEEP SEISMIC REFRACTION CROSS SECTIONS OF SAKHALIN (RUSSIA) ON THE DATA OF REINTERPRETATION USING 2-D INVERSION METHOD

PIIP V.B.

Department of Geology, Moscow State University, 119899 Vorobyovy Gory, Moscow, Russia

Abstract

Data of nine deep seismic refraction lines located in South Sakhalin and near areas were reinterpreted on the base of 2-D inhomogeneous model in the Moscow State University. The seismic works were carried out by Institute of Earth Physics and Sakhalin Integrated Institute in 1963 - 1964 years. Computer technology of refraction traveltime inversion by homogeneous function method was worked out in Moscow State University. This technology is automatic simple 2-D inversion of system of refraction traveltime curves to velocity field. Using this technology it is possible to calculate seismic cross sections containing velocity values and values of velocity gradient for every point of it, boundaries, slope and steep faults, folding zones and other features. Length of the lines varies from 250 to 750 km. Detailed cross sections were obtained down to 30-60 km depth. The new seismic cross sections correspond to old ones in depth of major boundaries and velocity values in average only. These studies support known tectonic model that active subduction zones "roll back" toward ocean. Remnants of oceanic plate and relic subduction zone are distinguished in the seismic cross sections and in the velocity maps of South Sakhalin. This relic subduction zone disposes in the rear of active Kuril arc. These cross sections were computed from observed seismic data without any preliminary proposals or initial models.

Keywords: Seismic, Refraction, 2-D inversion, Subduction, Sakhalin.

RELIC SUBDUCTION ZONE IN SOUTH SAKHALIN

Detailed seismic works along nine profiles were carried out in the South Sakhalin (fig. 1) in the period after the International Geophysical Year. For instance one of most detailed shooting geometry was yielded for profile 19 crossing south Sakhalin (fig. 2).Results of interpretation of received seismic materials were published in 1971 [1]. Now the first arrivals of refraction waves received in that period were inverted by 2-D inversion method using homogeneous functions approximation. More detailed information about this inversion method may be found in [2,1] and below in division '' Interpretation method''.

Here we examine the received cross sections. The main feature of seismic cross section for latitude profile 19 (fig. 3) which crosses exposures of ophiolitic rocks of South Sakhalin is the presence of relic subduction zone. Lower oceanic crust and mantle dip 20° westward from depth of 10 km (Okhotsk abyssal plain) down to depth of 20 km in the central part of profile forming a sequence of covers. Here oceanic slab is sloping and in the upper part of the cross section, the obducted block of oceanic slab is distinguished. At the depth of 30-50 km, relic oceanic plate dips about 20 ° again. In the west part of the profile, layered continental crust and mantle are present. Thick of the crust is about 35 km here. This architecture of cross section for profile 19 is confirmed by two cross sections for longitudinal profiles 18 and 20 that cross profile 19 and mis-tied at survey line intersections well.

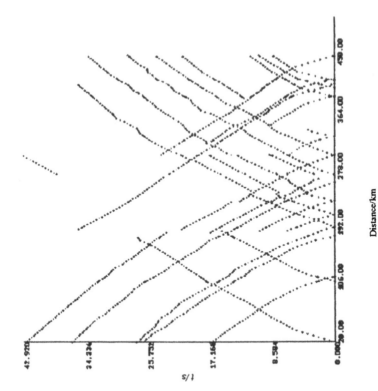

Figure 2. Refraction traveltime curves along profile 19 crossing the South Sakhalin. One of most detail shooting geometry was yielded along profile 19.

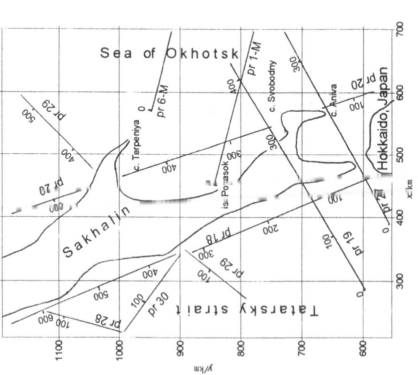

Figure 1.   Location map of seismic profiles in the South Sakhalin. Seismic works were carried out in the 1963 - 1964 years.

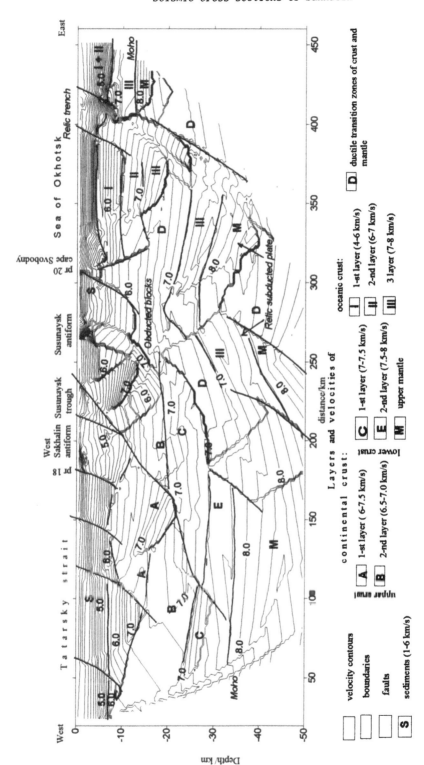

Figure 3. Seismic cross section and geological interpretation along profile 19 crossing South Sakhalin. Main feature of cross section is structure of relic subduction zone in the eastern part of the section. Contour interval is 0.2 km/s, vertical exaggeration is 3:1.

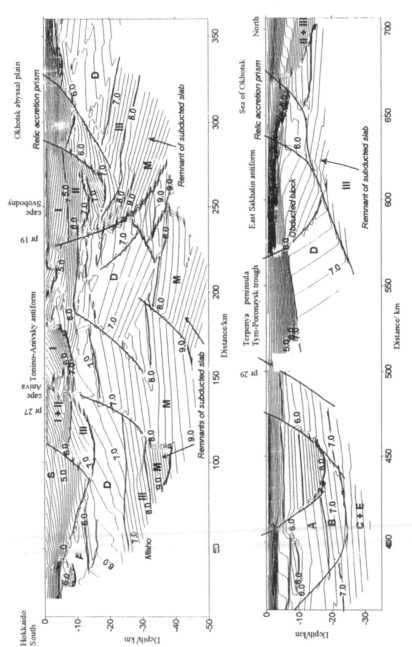

Figure 4. Seismic section and geological interpretation along profile 20 located near east shore of Sakhalin. Here the section has been divided in two halves: southern and northern parts. Main components of section are remnants of subducted oceanic plate distinguishing by high velocity and high gradient values. Contours interval is 0.2 km/s, vertical exaggeration is 2:1, symbols as in fig. 3.

Remnants of subducted oceanic plate are seen in the cross section of profile 20 (fig. 4) which disposes along east shore of Sakhalin. The remnant of oceanic slab located in the central part of cross-section contains thrusts and blocks of oceanic crust and mantle. One oceanic mantle thrust rises up to depth of 10 km and possibly served source of ophiolite at the surface. Relic accretion block can be seen at the 290 - 330 km of the profile. It is noticeable that remnants of oceanic plate show up sharply by high velocity (up to 10 km/s) and high velocity gradient values in the seismic cross section. It seemed that these remnants were intruded or embedded in the continental lithosphere that was disturbed by numerous faults here. In the north part of the seismic cross section anomalous crust is observed also. Relic subducted oceanic slab dips to south at 15 degrees here. Accretion prism may be seen also.

One of the remnants of oceanic plate reaches profile 18 located along west shore of the Sakhalin. Sharp restricted domain with increasing values of velocity and velocity gradient - vertical cross section of remnant of oceanic slab - is seen in the central part of seismic cross section (fig. 5). In the whole the lithosphere has continental type here. Crust thickness is about 40 km.

Locations of remnants of oceanic slab are seen well in the maps of seismic velocities constructed for depths: 8 km - average depth of crystalline basement, 22 km - of top of lower crust and for 32 km - average depth of Moho (fig. 6,7,8). In the construction of these maps the data of all of nine profiles were counted. Such velocity maps are similar to geological maps and tectonic schemes of crystalline basement, top of lower crust or Moho, because velocity does not change significantly along boundary surfaces and at the same time velocity values grow with depth as well as geological age. Thus low velocity area represents troughs and high velocity areas are rises in those maps. Examining these maps and cross sections together may make conclusion that the relic subducted plate consists of two segments here. Northern segment disposes parallel to the east shore of Sakhalin toward north of Terpeniya cape and southern segment is located parallel to the east shore of Sakhalin at the interval from Svobodny cape to Poyasok isthmus.

Location of relic trenches and obducted blocks are shown in the velocity map for depth of 8 km (fig. 6). Note that locations of obducted blocks have good correspondence with data of surface geology because the exposures of ophiolite complex rocks are known here. The two branches of relic subducted plates are distinguished in the map for depth of 22 km by relatively high velocities (fig. 7).

In the velocity map for depth of 32 km (fig. 8) the remnant of oceanic plate has form of tongue that leading from Okhotsk abyssal plain, reaches west shore of the Sakhalin and crosses it in the region of Poyasok isthmus. Northern segment of subducted slab, possibly was not shown as the net of profiles is infrequent. It is possible also that two branches of subducted slab joined together at this depth.

In the whole one of possible explanations of observed complex forms of cross sections of subducted plate by different profiles' planes is shown in the fig. 9. Intersections of the complex concave subducted plate by different vertical planes can produce such cross sections.

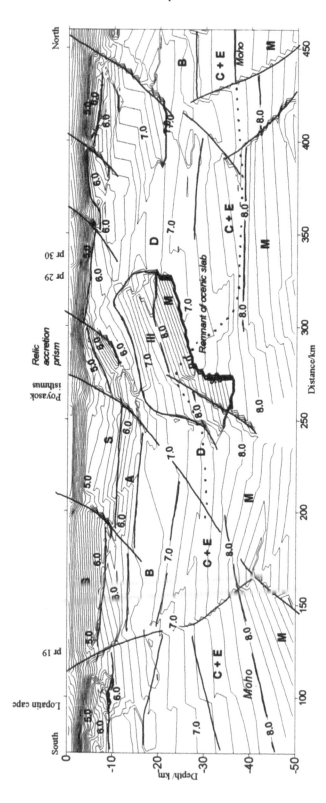

Figure 5. Seismic cross section and geological interpretation for the central part of profile 18 located along west shore of Sakhalin. In the centre of cross section the remnant of subducted slab is present. Location of Moho on the data of old interpretation is shown by dot line, other symbols as in fig. 3.. Contour interval is 0.2 km/s. Vertical exaggeration is 2.5:1.

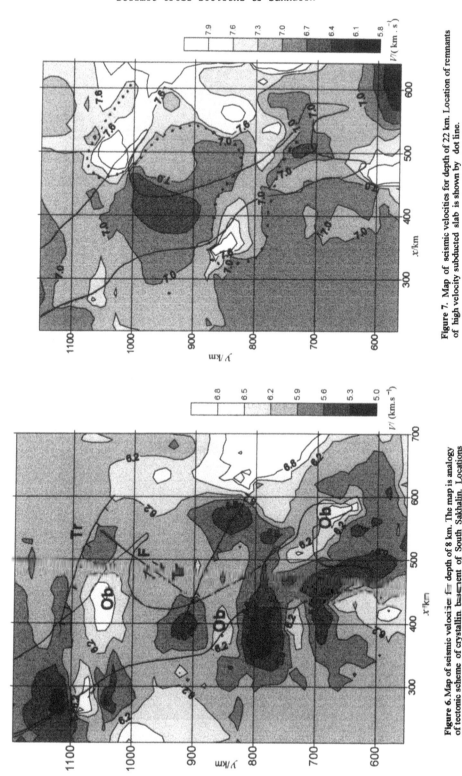

Figure 7. Map of seismic velocities for depth of 22 km. Location of remnants of high velocity subducted slab is shown by dot line.

Figure 6. Map of seismic velocities for depth of 8 km. The map is analogy of tectonic scheme of crystallin basement of South Sakhalin. Locations of relic trenches (Tr) and obducted blocks (Ob) of relic subduction zone and transform fault (F) are shown in the map.

Piip V. B.

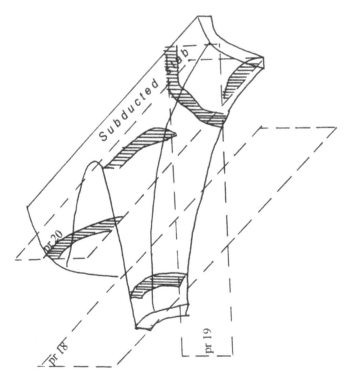

Figure 9. Speculative 3-D scheme of relic subducted slab in the South Sakhalin. In the scheme vertical planes of seismic cross sections are shown.

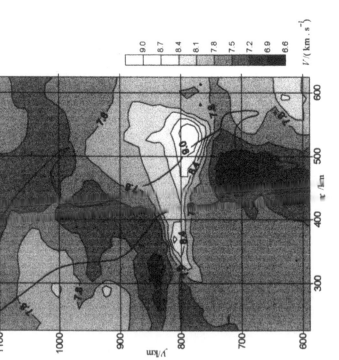

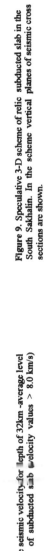

Figure 8. Map of the seismic velocity for depth of 32km -average level for Moho. Remnant of subducted slab (velocity values > 8.0 km/s) is seen on the map.

INTERPRETATION METHOD

Two dimensional inversion of refraction traveltime curves recorded at the surface of media may be fulfilled by methods of tomography using curve rays or by method of modelling - ray tracing. Both of these methods need a good initial model as corrections must be small. Problem of adequate division of model to blocks for local approximation that must correspond to observation system exists too.

Homogeneous functions technique for inversion of refraction traveltime curves is free from these problems. The method has been proposed and developed in Moscow State University. This technique is based on local approximation of real velocity fields by homogeneous functions of two coordinates [2] Approximate functions - homogeneous function of two coordinates - can describe inhomogeneous seismic media including seismic boundaries and unlimited gradients of wave velocities. Using local approximation of cross section by homogeneous functions, any velocity distribution in seismic cross section may be computed. Homogeneous functions are very suitable for approximation of real geological layered media.
2-D inverse kinematics seismic problem has been resolved on set of homogeneous functions described by formula

$$V = \left((x+c)^2 + y^2 \right)^{m/2} f\{\arctan[y/(x+c)]\}, \qquad (1)$$

here x, y is Cartesian coordinates of the cross section in kilometres. Next parameters of the velocity function (1) are computed in the inversion process:
m is the degree of homogeneous function (any real value);
c is parameter which defines a position of the local polar coordinates origin at the profile (calculations are produced in the polar coordinates);
$\varphi = \arctan \cdot (y/(x+c))$ is the polar angle;
$f(\varphi)$ is any increasing function of polar angle.
Investigation of eiconal equation for media where wave velocity is function (1) allowed to define that traveltime curves and seismic rays are characterised by relationships of similarity in such media. These similarity relationships are used for definition of local polar coordinates and degree of approximate homogeneous function for every pair of reverse traveltime curves along the seismic profile. It is proved also [2] that the 2-D direct and inverse seismic problems for media (1) can be transformed to 1-D problems for seismic media where wave velocity is function of polar angle only. This transformation is fulfilled before computation of function $f(\varphi)$ (1). Thus homogeneous function of any degree of two coordinates that approximates real velocity distribution, is computed for any pair of reverse traveltime curves from system of real traveltime curves along seismic profile. Set of curve seismic rays are calculated in the decision process also. Every such homogeneous function describes local velocity field. Algorithm of solution is stable and therefore it is possible to construct the final velocity cross section by superposition of local velocity fields in the common cross section plane. Error control is evaluated by difference of velocity values in adjusting points of different local velocity fields.

Computer program "GODOGRAF" for this interpretation process has been worked out and was used for interpretation of many field refraction data received in different regions and by different methods (engineering, mining, oil, deep and other refraction seismic) [3]. This is universal program for refraction interpretation. It is mportant to notice that this interpretation technique needs no initial model of cross section and so obtained cross sections are essentially objective. To use this inversion method, shooting geometry of refraction data may be any (minimum is traveltime curves from two sources). For complementation and interpolation of shooting geometry the special conversion of refraction traveltime curves to refraction time cross section is used. Traveltime curves are represented in next coordinates:

$$x = (x1 + x2)/2, \qquad l = \mathrm{abs}(x1 - x2)/2.$$

Here $x1$ is coordinate of receiver and $x2$ is coordinate of source, l is coordinate of the centre of receiver/source interval. The l-const contours are drawn. It allows to use program complex "GODOGRAF" for reinterpretation of old refraction data and also data with uncompleted acquision geometry.

Cross sections are formed by velocity contour lines. Contours are drawn with constant interval. Consequently the distance between contours is reverse proportional value with velocity gradient module. It means that we can automatically calculate complementary parameter of cross section - velocity gradient value. This parameter characterises velocity variation with depth and laterally. Seismic boundaries and faults appear in this cross section as areas of high contours density or areas of sharp change of contours density or contours pattern. Thus it is possible to distinguish at such cross sections not only different geological layers but slope faults, folded structures and local velocity anomalies.
Theory and algorithm of this interpretation process had been published in 1980-1990-s [2,4].

The inversion results were checked by direct problem solution many times, for instance [3]. The calculation of the ray paths and traveltimes through obtained cross sections were accomplished with the help of ray-tracing method using the tables calculated on rectangular grid 2-D velocity fields by program "GODOGRAF" as input data. Calculated times coincide rather well with the real ones. Rms value is appropriate with correctness of time determination. There were numerous of instances of such calculations and rms residual was inside correctness of time definition.

CONCLUSIONS

1. Using of two dimensional inversion method for old refraction data - method of homogeneous functions - allowed to discover complex features of deep structures of South Sakhalin.
2. Relic subduction zone and remnants of oceanic slab exist in the seismic cross sections of South Sakhalin in the rear of active Kuril arc.

REFERENCES

1. S.M. Zverev and Y.V. Tulina (Eds). *Deep Seismic Sounding of the Earth Crust of Sakhalin-Hokkaido seaside zone* (In Russian). Moscow , "Nauka", (1971).
2. V.B. Piip. New methods of interpreting of seismic time fields in media with variable velocities, *Moscow University Geology Bulletin.* **3**, 86-95 (1984).
3. V.B. Piip and E.A. Efimova. Karst and near-surface structure on the base of a new refraction interpretation. 6th international IAEG Congress/ 1990 Balkemia, Rotterdam. ISBN 90 6191 **1303**,1005-1008(1990).
4. V.B. Piip. Local reconstruction of seismic cross section on the refractive wave data on the base homogeneous function. (In Russian). *Fizika Zemli.* **10**, 24-32 (1991).

Proc. 30th Int'l Geol. Congr., Vol. 20, pp. 21-40
Liu (Ed.)
© VSP 1997

Deep Structure Pattern, Anisotropy and Continental Geodynamics Revealed by Geophysical Profiles and Transects in China·

TENG JIWEN[1], XIONG SHAOBAI[1], ZHANG ZHONGJIE[1], LIU HONGBING[1],
YIN ZHOUXUN[1], SUN KEZHONG[1], HU JIAFU[2], YANG DINGHUI[1], WAN ZHICHAO[1],
ZHANG BINGMING[1], ZHANG HUI[1]

1.Institute of Geophysics, Academia Sinica, Beijing 100101, China
2.Department of Geoscience, Yunnan University, Kunming 650091, China

Abstract

Based on the comprehensive studies on the profiles of deep seismic sounding, body and surface wave data from earthquakes, and the internal deep structure and geophysics field of 12 Global Geoscience Transect(GGT), the deep process and earth interior in China and adjacent region are studied, we also unveiled the zoning characteristics and layer-block structure of crust and mantle, lateral inhomogeneity and Anisotropy, special complex structure pattern, orogenic zone and sedimentary basin, continent extension and rift effect, earthquake activity and deep media environments of seismogenesis, detachment and deep and large fault, property of Moho, interface between lithosphere and asthenosphere, continental geodynamics and formation and development of intraplate and continental margin.

Keywords: Geophysical Profiles, Lateral Inhomogeneity, Anisotropy, Detachment, Asthenosphere, Rift Effect, Orogenic Zone, Continental Geodynamics.

INTRODUCTION

China is a territory where the tectonics of intraplate and plate margins are quit complicated. Under the combined effect of India plate, Eurasia plate, Pacific plate and Philippine plate, fracture structures were formed. Under the action of this unified force system, not only some developed fault systems formed in intraplate both in lateral and longitudinal directions, but also many blocks take these faults system as boundary(Fig.1).

In China, many profiles of deep seismic sounding(Fig.2a) and global geoscience transects(Fig.2b) provided highly accurate data for studying deep structure of crust and mantle, material composition and spatial distribution[1]. The observations for body wave and surface wave of natural earthquakes are valuable for approaching structure and property of mantle.

Based on the data of deep seismic sounding profiles and natural earthquake and full consideration to gravity field, geomagnetic field(aeromagnetic), electric field, thermal field, paleogeomagnetism and geotectonics as well as lithologic and chemical composition(such as mantle inclusion analysis), comprehensive research and geology interpretation were carried out in this paper under the direction of new theory. These results not only yielded new knowledge unable to reach in single discipline , but also unveiled zoning characteristics and complex structure of crust and mantle,

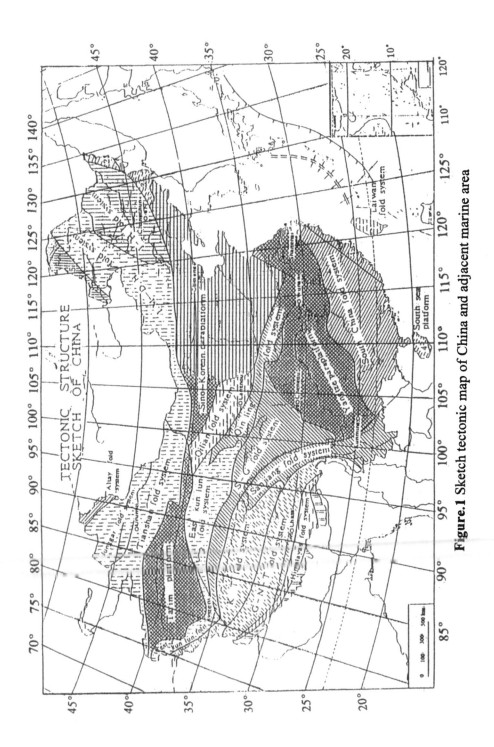

Figure.1 Sketch tectonic map of China and adjacent marine area

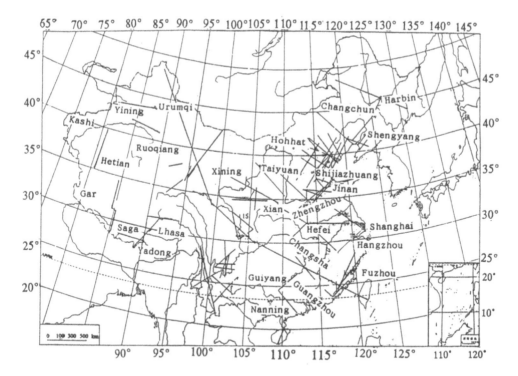

Figure 2a. Sketch map of deep seismic sounding profiles in China

lateral heterogeneity and anisotropy, the formation and evolution of earth's interior and geo-dynamics characteristics.

PROFILES OF DEEP SEISMIC SOUNDING AND THE GEOSCIENCE TRANSECTS IN CHINA

Seismic profiles obtained from deep seismic sounding are totally 50000 km long, those profiles passed through tectonic sutures, earthquake activity zones and mineral and energy resources zones. They are distributed mainly in the region of continental margin of east China.

Comprehensive Geoscience Transects in China are listed below:

A. The Yadong-Golmud Geophysical Transect; the length of the transect is about 800 km[2]. It passed through a series of geotectonic elements extending nearly east-west on the Qinghai-Xizang plateau, including hydrothermal areas (i.e. Yamzhoyiong lake, Yanbajin geothermal seismic region),and transition zones of compression and collision of Eurasia and India plates, earthquake active regions(i.e.Himalayan seismic belt, Damxung between India plate and Eurasia plate, etc.[3-7]): *B. The Xiangshui of Jiangsu-Mandulia of Inner Mongolia Geophysical Transect;* the length of the transect is 1476 km. It extends from southeast to northwest obliquely through the China-Korea platform and its margin, passing through eight terrains[8] of north Jiangsu, south Jiaonan, western Shandong, rift basin of north-China,Taihang mountain-Wutai mountain, Ordos basin, Hubhot-Baotou basin,Yinshan mountain and Inner-Mongolia fold system: *C. The east Dong Uzhum Xinqi of Inner-Mongolia-Donggou of Liaoning Geophysical Transect;* the length of it is about 1000 km. It extends east-west through the China-Korea craton, passing through a series of

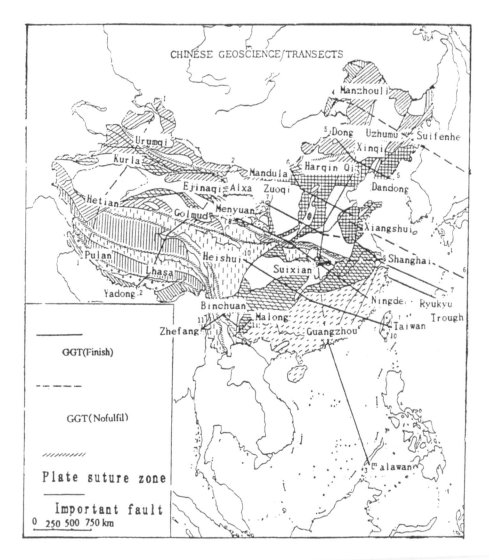

Figure 2b. The distribution of the comprehensive Geoscience Transects of China

geotectonic elements, such as Liaodong uplift, Xialiaohe downfault, Yanshan fold belt, Inner-Mongolia geophysical axis, Xilamulung river fold belt of the Inner-Mongolia-Daxin'anling fold system, and Hegenshan fold belt. And the south of the transect also crosses the Hacking seismic epicentre [9]: *D.The Fengxian of Shanghai-Alxa Zuoqi of Inner-Mongolia Geophysical Transect*;Its length is about 1500 km. Its western end starts from the eastern boundary of Alxa block, the China-Korea craton, extending SEE through the west margin fold belt of Ordos and its indentation, the Shanxi faulted uplift, south of north China indentation and Xuzhun fault uplift, then to the north Jiangsu-south Shandong terrain and the north Jiangsu fault depression or south Jiangsu fold-belt of the Yangtze craton [10]: *E.The Suizhou of Hubei-HaidinQi of Inner-Mongolia Geophysical Transect;* the transect crosses over the geotectonic element of the Qinlin fold belt-Tongbai-Dabie mountain, east Taihang mountain, north-China and so on[11]: *F.The*

Menyuan of Qinghai-Ninde of Fujian Geophysical Transect; the length of the transect is 2200 km.From northwest to southwest, it passes through 5 geotectonic elements of the Qilian fold system,China-Korea block, Qinlin-Dabie orogenic zone, Yangtze craton,and south-China orogenic zone[12]: *G.The Yunnan's Zhefang-Malong Geophysical Transect;* the transect is 1800 km long. It passes through the Yangtze craton, the orogenic zones of Sanjiang Rivers, and the southeast part of the Gangdise-Nqanqentanglha orogenic zone[13]: *H.The Tianchang of the Xia Yangtze region-Shanghai Geophysical Transect(HQ-13);* surveys of seismic broad-line reflection[14], seismic broad-angel reflection and seismic deep refraction are carried out along the transect: *I.The Heishui-Quanzhou Geophysical Transect;* the length of the transect is 4600 km. From west to east, it passes through the Sichuan basin, Bayanhar mountain, Yangtze craton,Xuefeng mountain,Luoxiao mountain, and Fujian and Taiwan tectonic zone[15]: *J.The Guangzhou-Balawang Geophysical Transect;* the transect stretches over the whole South China Sea. It is a comprehensive transect with 1500 km in length which can link the sea and land of China[16]: *K.The Manzhouli-Shuifen River-Sea of Japan Geophysical Transect;* the length of the transect is 2100 km[17]. It starts from Russia on west, and reaches the Sea of Japan no the east. It has great significance to link the Eurasia plate with the Pacific plate. The transect passes through the geotectonic elements of Siberian, Mongolia-margin of Sea of OchOtsic sea, sea of Bajikal rift, Mongolia-Xing'an orogenic zone, the east-west margin belt,the Songliao basin,and Zhangguancai mountain, etc..: *L.The East China Sea geophysical transects;* the west endward of the transects starts from the offshore region of China mainland, extending east to the Philippine Sea in the east of Ryukyu oceanic trench. Its total length is 700 km or more[16].

THE LATERAL INHOMOGENEITY AND COMPLEXITY OF CRUST AND UPPER MANTLE IN THE CONTINENT OF CHINA

The stratified characteristics, low velocity layer and high velocity gradient interlayer in the crust.

In north China, there are two low velocity layers in crust of Qinghai-Xizang and all earthquakes occurred above the low velocity layer. however, shallow earthquakes occurred above the 1st low velocity layer(upper crust). The 1st low velocity existed in Qinghai-Xizang whether in north (Yandong-Naqu-Siling Co) or south Tibet (Puma lake-Dinggye-Peiku lake), all of them extend to west and thin out at Bange and Dingye[18]. The researches on deep process in different regions indicate that low velocity layers developed well in shallow crust of seismic, geothermal and tectonically active regions. For example, in Quanzhou area in continental margin of southeast China, the burial depth of low velocity layer is 13-14 km[19](Fig.3), but the temperature is 120°C at 252 m depth in the 1st well in Longwi Normal School, Zhangzhou region. There is a low velocity body of large-scale below 9 km and it extends 100 km in lateral with 10 km of thickness and 5.9 km/s of average P-wave velocity. From the fact of P wave velocity decrease and disappearance of S wave reflection energy we deduced that partial melting body exist in shallow crust.

New understanding of Moho

The understandings of Moho are deepening with increasing of deep explosion seismic data. Over a long term, Moho is considered as sharp discontinuity, but this view has not a universal significance now. Because the explosion seismic data show that Moho is a complex transition or variation zone of low and high velocity layers interbeded, its thickness is kilometres. Comparisons between seismic reflection data from active orogenic zone and pale-orogenic zone indicate the Moho difference. Unlike pale-orogenic zone, such as Aba and New England fold zone, Taihang mountain and its neighbouring area, Himalayan mountain systems are the youngest orogenic zone, tectonic extremely active, the mountain body is highly uplifted, but no mountain root is formed in this region, the crust is not very thick, gravity is not isostatic and it is still uplifting. Moho is a bunch of thin layers that is extremely inhomogeneous in lateral with large dislocation and there are large nappe structures in crust.

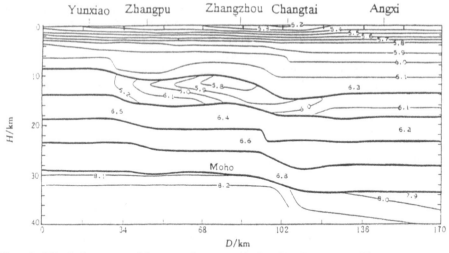

Figure 3. 2-D velocity structure of the crust and upper mantle along Yunxiao - Anxi profile

In addition, such as in Caledonian orogenic zone, the bottom of wide tilt reflection surface of low crust show obviously near horizontal reflection of Moho, this indicates that Moho was formed later than orogenic zone. Therefore, based on the flat shape of Moho under pale-orogenic zone, it can be considered as a relatively younger tectonic zone. In Qaitamu region of Qinghai province, Manzhouli-Suifenghe region and the north part of Qinghai-Xizang plateau, the seismic profiles of deep reflection and wide-angle refraction survey show clearly that Moho discontinuity is composed of thin layers, it was attributed to the intrusion of late period magma or phase change as well as deep material differentiation and isostatic affection. The transition zones of crust and mantle exist universally in the middle Hebei, western Shandong uplift zone and Taihang mountain fault. The intrusion of magma at the bottom of crust and the extension process into crust of magma may play an important role[22,23].

The boundary between lithosphere and asthenosphere
Lithosphere is a hard shell of Earth, but the asthenosphere is a hotter, softer and plastic rock layer. Their division is mainly based on the following essential factors, namely, the difference of temperature, the travel velocity of seismic wave and Q value, and the difference in rigidity. The burial depth of mantle high conduction layer can be determined by magnetotelleric sounding, the depth of upper-mantle partial melting zone can be determined by average thermal value, the burial depth of asthenosphere also can be determined by P-T condition of inclusion and stress level.

Based on above-mentioned method, a dividing boundary of lithosphere and asthenosphere is given and the velocity structure and spatial distribution are studied by many Global Geoscience Transects. North China plain-East Sea of China-Island arc of Ryukyu-the Philippines Sea-Mariana Trench-Pacific Ocean Geoscience Transects unveil that the top of asthenosphere is 96 km in south China continental margin, 35-40 km in deep sea basin area, 50-70 km in south part of South China Sea and north Sulu sea area. Obviously, from basin, margin, near sea, to sea area and deep basin area, the burial depth of asthenosphere has been determined by comprehensive geology survey and geophysical research (such as seismology, gravity, heat flow, drag-net sample along sea bed), the thickness of lithosphere in north China plain and offshore area is 60-80 km, in Philippine sea is 50-70 km, in Paresabeila sea basin is 30 km, in Maliane trench is 20 km. This profile also reflected the relationship between age and thickness of lithosphere, lithosphere is thin under younger ocean shell and it is thick under older continental shell. In intraplate area, the thickness of lithosphere in north China plain and offshore area is 80±2 km and 70±5 km respectively[26].

The crust is very thick (70-78 km), but the burial depth of asthenosphere is rather shallow(90-100 km)[27] in Qinghai-Xizang plateau, so the special structure of thick crust and thin lithosphere formed in this region. In northwest orogenic zone and basin area it is 100-120 km[28], it is uplifted in basin area and descendant in orogenic zone, the difference may reach 5-10 km.

In general, the thickness of lithosphere increases from east to west, Daxing'anling-Taihang Mountain-Wuyi Mountain is a salient variation area(thickening area) of lithosphere thickness. North-South tectonic zone is a dividing zone between east and west and from shallow to deep. The thickness of lithosphere is thinner obviously beneath Neozoic rift.

The crust thins from west to east in China. In east China, it lies in terracing distribution and was intersected by a series deep and large faults that cross Moho and reach to the top of upper mantle. Due to lower velocity layer been intersected, these faults can be deduced as resulted in Mesozoic or Neozoic. From the profile of crust from Hindukush in west via Xizang plateau, Qinghai to east sea, the burial depth of the top of upper mantle in Hindukush area is 60-65 km and it increases gradually from west to east, reaching 75 km beneath the Xizang plateau, then it decreases quickly from Xizang plateau to east, it is 30 km in east sea area. The highest Qb value is 200-500 in upper and middle mantle-low crust crystallized basement.

DEEP INTERNAL PROCESSES AND DEEP MEDIA ENVIRONMENT OF EARTHQUAKE ACTIVITY

Earthquake activities show obvious zoned characteristics in the mainland of China and adjacent area, and most of them are shallow earthquakes(10-30 km). A group of fault with some local dislocation were discovered in epicenter area under Xingtai Earthquake zone, it's horizontal displacement is 2-3 km. The faults are disconnected with others, but deep faults reaching to top of upper mantle were found under them. The epicenter area is characterized by high absorbing zone(low Q value), locally uplifted upper mantle zone and variation of geophysical field[22]. From the seismic reflection profile, there are listric faults in upper crust(Fig.5), such as the Xinghe faults extending to 8 km depth[29] The fracture in Xingtai main-shock zone is a strike slip fault, its dip is about 80 degrees, the foced depth is 9 km, so the seismogenic fault is a normal-fault, it was interred that a deep fault from focus area to Moho existed, this fault may be seismogenic fault of Xingtai large earthquake. Due to the intrusion of deep hot material, expansion with different degree occurred in crust and additional thermal stress produced, so it can be deduced that the formation of north China basin and occurrence of Xingtai earthquake were caused by hot material upwelling along the fault channel from upper mantle.

The tectonic environment and mechanism of earthquake generation and development are very complex, so detailed analysis on concrete physics and mechanics process must be carried out on each earthquake. By means of the deep structure studies on Dongwang earthquake[30], Tangshan earthquake[31,32], Haicheng earthquake[40], Linfen earthquake[33], Bohai earthquake and Xichang earthquake, a comprehension concept about deep media environment and deep internal process of seismogensis were obtained.

Earthquake activities in China and adjacent area are closely related to the structure of crust and upper mantle, and characterized obviously by belt patterns. Such as in Hindukush earthquake belt and Xizang plateau earthquake belt, not only earthquakes occurred frequently, but also the focal depths reach to upper mantle(as shallow and middle depth earthquake), but they are shallow earthquakes in north-south seismic belt. Earthquakes are rare in south China mainland, some earthquakes are distributed in the south of continental margin. These earthquakes are centralized in middle and upper crust in the viewpoint of Q value.

EVIDENCE OF DECOLLEMENT IN LITHOSPHERE

Large nappe-decollement structures in Himalayan mountain
Based on deep seismic sounding and quasi-vertical reflection, there exist two low velocity layers beneath Yarlung Zanbo river and the both its north and south sides and its adjacent area. Both low velocity layers uplift gradually to south, the upper low velocity layer is connected with the

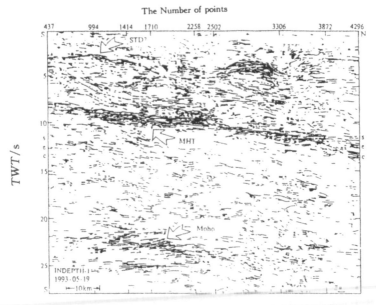

Figure 4. Vertical reflection profile of INDEPTH-1 SDT Southern Xizangan Detachment, MHT-Main Himalayan Thrust which corresponding to the upper boundary of Indian crust.

main boundary large fault and formed large decollement extending from south to north in this area(Figure.4)[34,35]. The uplift of Himalayan mountain, intense activity of transition belt of collision and compression are closely related to the compression and collision between two continental plate and Indian plates wedged into Xizangan crust toward north.

Large decollement in north China basin
From Fig.5, the profile shows two downfaulted basins, the east basin belongs to Shulu downfaulted basin which corresponds to the Xinghe fault, the west basin is Jinxian downfaulted basin, its main boundary is Jinxian fault. The region between two basins is quasi-translucent zone

of seismic reflect field or weak reflection zone formed by little energy discontinuously short reflection segment[23].

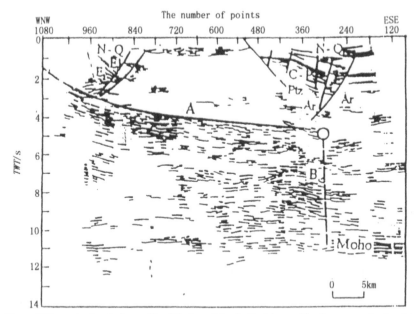

Figure 5. Deep seismic reflection profile of Lincheng-Juhu. A--Stacking profile of common depth point; B--Reflection interpretation profile

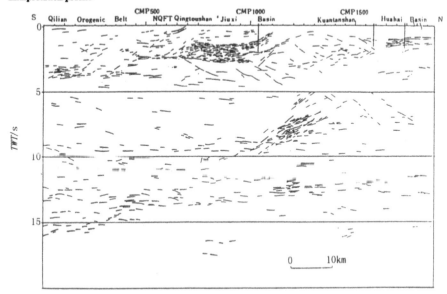

Figure 6. Common depth point reflection profile in Diaodabun Mt. -Huahai basin

Large detachment from north to south in the crust structure of western area
It can be found from Fig.6 that reflection event A in middle profile is decollement and it is

represented as a multi-cycle reflection belt and its width is about 1.0s[36]. From north end 2.5s to south, reflection event can be tracked about 1s. According to thetendency of decollement on the profile, it can be inferred that the decollement will emerge to surface of earth at north end of Huahai basin profile. Reflection event of decollement is characterized by continuity and strong energy etc., and it continued about 1.0s in time section. The dip of the reflection belt at north end of profile is 30°, and it decreases to south gradually, it is 10° under middle Jiuxi basin, but it decreases to 5° to the north.

ANISOTROPY OF VELOCITY DISTRIBUTION IN EARTH INTERIOR

Anisotropy reflected by difference of velocity of seismic surface wave.
(1)The difference between Vsv and Vsh all occurred in Qinghai-Xizang plateau, northwest orogenic zone and sedimentary basin[37]. $Vsv>Vsh$ and characteristics of horizontal compression appear under depth 100 km of Qinghai-Xizang plateau.

(2)North-South tectonic zone. A block diagram of Vsv-Vsh at different depth[38] is given. It is found obviously that the difference between Vsv and Vsh is very small, but Vsh is larger than Vsv as the depth increased.Anisotropy strengthens low velocity layer of mantle, but it is not obvious in north transect.

Polarization of SKS phase and anisotropy
(1)Kunmin of Yunnan area, faster polarization orientation of SKS wave is NE-SW, faster direction of Pn wave is NW-SE. The direction of quickest velocity of Pn wave is identical with the direction of Kangdian lithosphere block movement. Split time of SKS is 0.58s, it was decreased to 0.35s in upper crust and top of upper mantle[39], The decrease of split time of 0.3s show that SKS split come from upper mantle and anisotropy rate of Pn wave is about 3.3%. This was caused by the collision between Indian plate and Eurasia plate, which makes Kangdian block move to southeast direction, and produce the dominant orientation of olivine crystal axis[Fig.7].

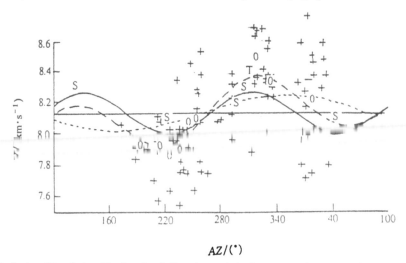

Figure 7. Distribution of Pn velocity with orientations in Kunmin area. "+"--observatory value; "O"--zoning average value of orientation; "T" and "s" are quoted data; Horizontal solid line--total average value; Dot line--variation with single orientational angle; Curve--variation with dual orientation real angle; Dashed line--curve which obtained from fitting measurement

(2) In area of Qinghai-Xizang plateau, there is a great difference in anisotropy on the south and north of Yarlung Zangbo River . The average polarization directions of faster wave are about N70°E in the area north of the river(Fig.8) and it increases from north to south. It is close to E-W near Yarlung Zangbo River and it is about N25°W at the area south of the Yarlung Zangbo River. In addition, there is a difference in strength of anisotropy[40], travel time difference between

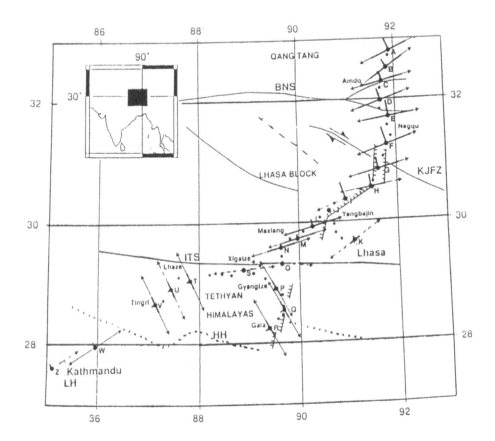

Figure 8. Tectonic structure and anisotropy of media under each seismic station in Xizang plate. Length of short line show the difference of the strength of anisotropy. Short lines show direction

faster and slower wave reaches 1s near Naqu at north region of Yarlung Zangbo River and local variation exits, but the anisotropy is weak and stable in the area at south of the river and δt=0.2s. This indicates that the direction of pressure of India plate is toward northeast on the north side of Yarlung Zangbo River. The difference in anisotropy in the areas north and south of Yarlung Zangbo River also proves that the subduction leading edge of India plate does not pass across Yarlung Zangbo River.

Anisotropy of lithospheric orientation in China mainland
56 SKS phases were distinguished clearly from broadband seismological record by 8 seismic observation stations[41]. After analysis and inversion, it was found that S wave split exists obviously with delay time between 0.7s to 1.7s(Fig.9). Anisotropy of top mantle is the main cause

which leads to the S wave split when travelling through the mantle(SKS phase) and it can be interpreted by crystal dominant arrangement of mineral in mantle.

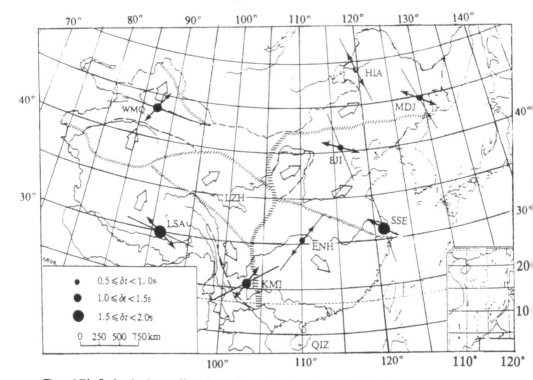

Figure.9 Distribution sketch map of inversion results on orientation anisotropy of lithosphere in China mainland. Thick line arrow: faster axial direction; Block dot: time retard(delay); Tilt line: boundary of intraplate; Thin long line: faster axes of shear wave inferred from deformation status of intraplate

Should be point out, due to strain lead to dominant arrangement of crystal lattice, elastic rate of rock media characterize anisotropy. If we propose 50%-70% of upper mantle rock is olivine and the directions of crystal axes of 15% olivine identical to each other, then one second of delay are equal to travelling distance of seismic wave in anisotropic media of 115 km[41,42]. Therefore, this result implies that the thickness of anisotropic media layer which leads to S wave split is about 80-190 km. These numeral ranges are corresponding to the thickness of lithosphere under China mainland inferred by low velocity layer high conductivity layer in upper mantle.

RESEARCH ON THE FORMATION, EVOLUTION AND DYNAMICS OF LITHOSPHERE

Special tectonic background of Qinghai-Tibet plateau collision orogenic zone

From profiles and transects of crust in north China and Qinghai-Xizang plateau,we can find complex layer-block tectonics in crust and upper mantle, which are transition-zone consisted of thin layer bunch or gradient layer. The tectonic of Himalayan collision orogenic zone are more complex, there exist embedding model in lithosphere, thickening and uplifting of crust in Qinghai-Xizang plateau are caused by effect of compression. The upper and lower surface of this "wedge" are upper and lower low velocity layer in crust, but the lower boundary surface of low velocity layer of crust is Moho zone at the area south of Yarlung Zangbo River, and it was end by hindering of Yarlung Zangbo high-angle fault[5,6,8]. The difference of anisotropy along both sides south and north of Yarlung Zangbo River etc. all show that this"wedge" does not pass over Yarlung Zangbo River. Collision and compression transition zone are consisted of a series of nonlinear stratified, uniform and short transect of seismic zone, but bury depth of Moho boundary belt are worth bearing in detailed discussion. Because the media are broken heavily in this area, it may be caused by complex structure pattern which was a result of upper and lower disorder of fracture and fault.

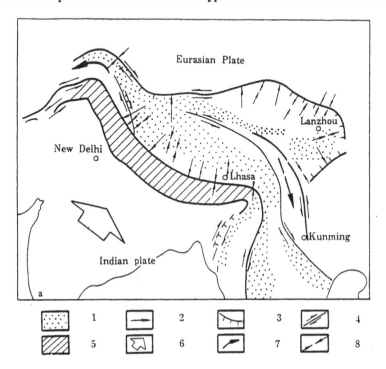

Figure 10. Simplified map of the transitional belt of collision and compression of plates and deep material flow toward southeast in Qinghai-Xizang plate. 1.Deformation of compression; 2.Direction of stress; 3.Overthrust; 4.Strike slip fault; 5.Transitional belt of collision; 6.Direction of movement of plates; 7.Direction of dominant and compression between Indian plate and Eurasian plate movement for crustal and mantle matter; 8.Direction of principal stress axis.

The last study of wide-angle seismic reflection data show that the suture belt of Banggonghe-Nujiang incline to north and runs down to upper mantle, on its south side there is a deep depression of Moho, Gangdise block underthrusting to north and Qiangtang block overriding passively to south caused 10 km jumping at the bottom of crust.

Qinghai-Xizang plateau uplifting, crust shortening and thickening are related to deep process of Himalayan collision orogenic zone. They are results of collision and compression between India plate and Eurasian plate and dual direction action from north and south, and the evolutionary pattern formed since Paleozoic era. Should been seen, not only differentiation and adjustment of mater but lateral flowing was taken place during the process of continental plate collision(Fig.10).

Deep internal processes of Panxi tectonic zone

The Panzihua-Xichang tectonic zone is effected by the tectonic movement and deep matter migration of all blocks around it, so a passive "re-activated" ancient rift was formed in this zone[43]. It is characterized special geophysical field and internal deep processes, and it is an active zone of modern tectonic movement. Q values are low in this area, and it can be divided into three zones(Fig.11). Such as western part: Zhongdian, Ninglang, Yongsheng, Huaping etc. Q value is 386-435,eastern part; Zhaotong, Qiaojia, Dongchuan etc., the Q value is 480-616,middle part; Mianning, Muli, Zhaojue, Puge, Yanbian, Miyi, Huili, Panzihua, Pingdi etc., Q value is about 251-330. Additional, the velocity structure and material property are difference between western and eastern part. These results show that Kangdian tectonic axial zone are low Q value area, The Q value in western area is higher than in middle area, but eastern area is higher Q value area, the main reason of Q value low in middle part and occurrence of structure difference is that the comprehension action of thermal matter migrated in deep crust, moving crust and large fault system.

From Fig.12, The burial depth of mantle asthenosphere is 80-90 km in this area, mantle flow at bottom of lithosphere still upwelling, heat flow value are high in axial part, depth focus in flank is larger than in axial part, uppercrust is very inhomogeneous , clear reflection fail to acquire in this area, velocity is lower in top of upper mantle (7.6-7.8km/s). Since basalt erupts in Permian, this region experienced many tectonic movement and magmatic activity, not only formed many mineral resources collections, strong earth-quake activity, but also made up a special rift type and modern complex tectonic pattern, so this a passive "reactivated" ancient rift, this replenishes the classification of the rift in the world. Should been appoint, rift tec-tonics of China have it's own special features.

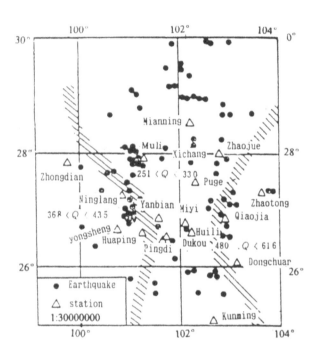

Figure 11. Distribution characteristics of Q value in Kangdian tectonic zone and its adjacent area(shadow area express difference boundary belt of Q value distribution)

Tanchong-lujiang fault system

It is a deep and large fault system which extended to the top of upper mantle. Structure of lithosphere is different obviously in both side of this fault, reflection from Moho under the fault is not clear and 2-3km vertical magnitude of fault were found. Seismic reflection and refraction show that Tanchong-lujiang fault system are shear system which are consisted of 3-4 faults[15,44,45](Fig.13). Lower crust in eastern part of Tanchong-lujiang fault system underthrust to west show that there is a compression subduction zone in deep mantle material upwelling in local areas which take this fault system as channel and vertical activities exist deep.

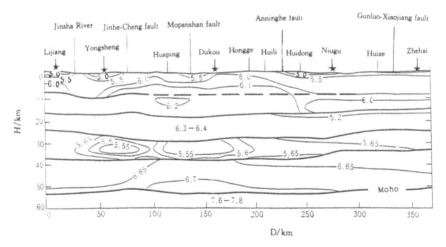

Figure 12. Two-dimensional velocity structure profile of Panxi tectonic zone

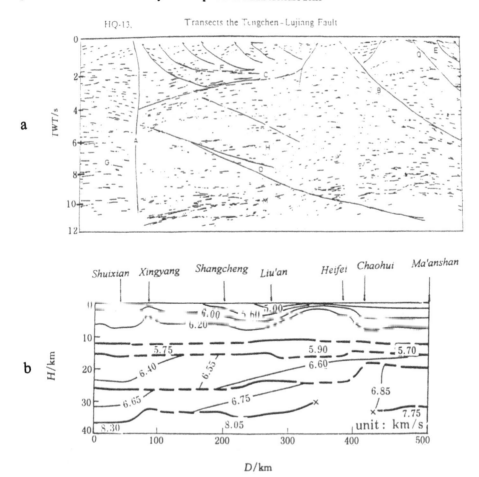

Figure 13. Distribution map of velocity structure section in Tancbong-lujiang fault system[15]. a—Seismic reflection dual travel time profile; b—Two dimension velocity structure section[44]

Dynamic characteristics of China continent

Obviously, eastern part and marine area of China just situated at special belt of lithosphere structure and geophysics field. Thickness of crust thinned from west to east, and it can be divided to some zoning boundary[30,46] (Fig.14). Under the action of three plates, continent of China and

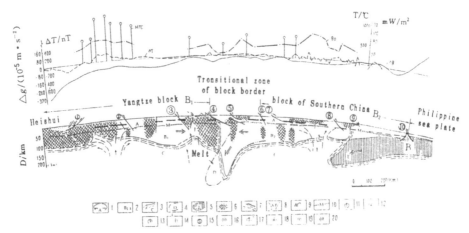

Figure 14. The interpretational section of southern China intraplate. 1.Crust and low velocity layer; 2.Lithosphere; 3.Asthenosphere; 4.Subduction zone for block of south China; 5.Subduction of Philippine Sea plate; 6.Rigid block; 7.Fault and ductility shear zone; 8.Motion direction of mantle thermal material; 9.Motion direction of plate; 10.Moho; 11.Longmen Mt. fault zone; 12.Hugyin Mt. fault zone; 13.Songtao deep fault zone; 14.Thrust deep fault zone of Xupu collision border; 15.Jinlan buried fault zone; 16.Calin-Chen xian fault zone; 17.Suichuan fault zone; 18.Zhengbe-Dabu deep fault zone; 19.Changle-Nan'an fault zone; 20.Seacoast.

its surrounding area becomes broken and formed the tectonic pattern of diversity blocks inserting together(Fig.15). From earth surface to deep part of mantle . Obviously, tectonic pattern of continental margin area of east China was controlled by global tectonic movement, it was put together by several block and was transformed and compressed strongly. Such as continent boundary of north part of South China Sea formed by old land, collision broclcen belt, driving continental and passive margin formed after expand of continent margin. Therefore, whether modernly tectonic pattern in Taiwan area, or formation of trench, arc and basin system, all experienced a series of complex evolutionary process, and a series of sedimentary basin with different genesis formed in continental margin and arc in South China.

Due to cooperation action among Philippine Sea block, Pacific block and India block, local block not only wedging into Qinghai-Xizang plateau and continental in east China, but also pushing to inland. To sum up, following conclusion were obtained [5, 6, 26, 30].

(1).Continental margin area was pressed by western movement of Pacific plate and Philippine Sea plate, so the uplifting and deformation occurred and a series of orogenic zone and basin formed in this area. Strong shear slip, temperature upwards and partial melting occurred and valance and a series of granite formed in crust.

(2).Yangtze craton crust at east part of Tanchong-Lujiang fault underthrust to north along this fault, as a result, tensile-uncouple rift or downfaulted basin occurred at west side, it also leads to late Jurassic Opoch-Cretaceous. During tensile-uncouple apart process, crust extension, mantle uplifting and middle crust thinned.

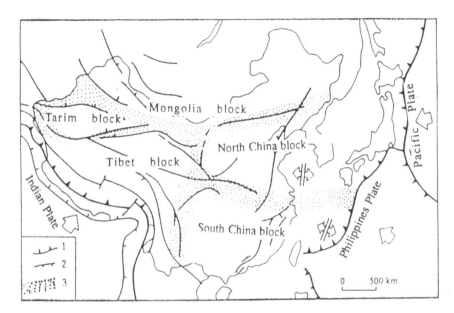

Figure 15. Distribution map of intraplate(block) tectonics in China and adjacent marine area

(3).Along Daxinganling-Taihang Mountain-Wuyi Mountain-Miaoling Mountain, lithosphere thicken, upper mantle was disturbed strongly, mantle material migrated upward lead to magmatic intrusion and volcanic eruption occurred, and large scale NNE direction gravity gradient belt formed, and series fold, overthrust and shear separating occurred in this area.

(4).North and south tectonic zones are result of cooperation action among east part of north China, Yangtze craton and Qinghai-Xizang plateau. After collision between India plate and Eurasian plate, stress did not drain to end, so it transmitted to north, deep material divergence, adjustment and flux lateral at occurred again. Not only lead to a series of NW-NNW strike slip fault formation, but also migration of material flow turn to south hindered by Yangtze craton, and some rhombi block with load boundary formed. In this area, over flow basalt which is symmetrical to axial erupted, upper mantle uplift, earthquake occurred frequently, hydrothermal activity intensively, and plentiful metal and nonmetal minerals source formed in this NS direction rift zone.

(5).Collision and compression between India plate and Eurasian plate not only lead to Qinghai-Xizang plate thickened, intensive activity of earthquake and hydrothermal, strong linear aeromagnetic anomaly belt along Yarlung Zangbo River, Ophiolit suite and mixed pile up extend to 1500 km, but also result in the enormously thick crust and relatively thin lithosphere[4,6,35]. In Himalayan mountain area, gravity do not isostatic, mountain system still rising, and the motion and collision of two continental plates from the roof ridge of the world and transition belt of collision and compression.

(6).To west again, it is Hidukush seismic zone. In this zone, thickness of crust is 50-60 km, depth of earthquake source is middle, energy and frequency is high, source plane distributor as a shape of V and the collision between India plate and Eurasian plate first occurred here, then India plate play an anticlockwise rotation and collided with Eurasian plate in India-Burma mountain area. Earthquake with middle source depth also occurred in this area. These are east-end and west end (arc top) of Himalayan arc mountain system, ancient Mediterranean sea disappeared by compression and modernly tectonic pattern was formed. So we can say, under the cooperation action of India plate, Pacific plate, Philippine plate and Eurasian plate, China continent and marine area are insetting structure which set by some broken block.

REFERENCE

1. Teng Jiwen. The seismic study of lithosphere physics in China(in Chinese), *Acta Geophysica Sinica*, 37, Suppl., 1—14, (1994).
2. Wu Gohgjian, Xiao Xuchang, Li Jiangdong et al. Reep track of world advances, explore for new research fields-an introduction to the "Yadong-Golmud Geoscience Transect" in Qinghai-Xizang Plateau(in Chinese), *Bulletin of the Chinese Academy of Geological Science*, 21, 1-8, Beijing: Geological Press,(1990).
3. Teng Jiwen,Wang Shaozhou, Yao Zhengxin et al. Characteristics of the geophysical field and plate tectonics of the Qinghai-Xizang plate and its neighboring region(in Chinese), *Acta Geophysica Sinica*, 23, 254—278,(1983).
4. Xiong Shaobai,Teng Jiwen.. The thickness of the crust and undulation of discontinuity in Xizang(Tibet) plate(in Chinese), *Acta Geophysica Sinica*, 28, Suppl., 16—26,(1985).
5. Teng Jiwen, Xiong Shaobai, Yin Zhouxun. Structure of the crust and upper mantle and velocity distribution characteristics at northern region of the Himalayan mountain(in Chinese), *Acta Geophysica Sinica*, 26, 525—540, (1983).
6. Wu Gongjiang, Xiao Xuchang, Li Tingdon et al. Lithospheric structure and evolution of the Tibetan plateau, the Yadong-Golmud geoscience transect, *Tectonophysics*, 219, 213 —221, (1993).
7. Zseng Jungsheag, Zhu Jieshou. Three-dimensional structure of seismic wave velocity and continental collision model beneath Qinghai-Xizang and its eastern part area(in Chinese), *Acta Geophysica Sinica*, 14, Suppl., 523—533, (1992).
8. Ma Xingyuan, Liu Changquan Liu Guodong. *The Xiangshui of Jiangsui-Mongolia-Donggou of Liaoning Geoscience Transect*(Explain book)(in Chinese), Beijing: Geological Press, (1992).
9. Lu Zaoxun,Xia Huaikuan.*The Dony Uzhom Zinqi of inner-Mongolia-Dongou of Liaoning Geoscience Transect*(Explain Book)(in Chinese), Beijing: Geological Press, (1992).
10. Sun Wucheng, Xu Jie Yang Zu'en et al. *The Fengxian of Shanghai-Alxa Zuaqi Geoscience Transect* (Explain Book)(in Chinese), Beijing: Geological Press, (1992).
11. Sun Wucheng, Ma Baolin, Song Sonyan et al.. *The Suizhou of Hubei-Harqin Qi of Inner-Mongolia Geoscience Transect*(Explain Book)(in Chinese), Beijing: Geological Press, (1992).
12. Lin Zhongyang,Cai Wenbai, Zheng Xiebo et al.. *The Menyuan of Qinghai-Ninde of Fujian Geoscience Transect*(Explain Book)(in Chinese), Beijing: Geological Press, (1992).
13. Kan Rongju,Hanyuan.*The Yunnan's Zhefang-Malong Geoscience Transect (Explain Book)*, Beijing, Geological Press,(1992).
14. Weng Shije,Chen Hushen. Deep seismic probing of continental crust in the lower Yangtze region, Eastern China, *Tectonophysics*,173,297—305, (1990).
15. Yang Guangceng, Xu Mingcai Tang Wenban et al.. Eastern Qinling seismic reflection profile(in Chinese), *Acta Geo phywica Sinica*, 37, 749—758, (1994).
16. Liu Guangding. *Geological-geophysics features of China seas and adjacent regions*, Beijing: Science Press, (1992).
17. Yang Baojun. Study of vertical reflection seismic along Marzhouli-Suifenghe Geoscience Section(in Chinese), *Annual of China Geophysical Society*, 246, Beijing: Seismological Press, (1993).
18. Liu Hongbing, Teng Jiwen, Yin Zhouxun et al. Characteristics of the 2-D crustal structure and fault distribution beneath the Siling Co-Yangamdo in the northern Xizang plateau(in Chinese), *Bulletin of the Chinese Academy of Geological Sciences*, 21, 250—259, (1990).
19. Xiong Shaobai,Jin Dongmin,Sun Kezhong et al.. Some characteristics of deep structure of the Zhangzhou geothermal field and its neighborhood in the Fujian province(in Chinese), *Acta Geophysica Sinica*, 34, 55—63,(1991).
20. Tseng Jungsheng, Teng Jiwen Kan Rongji et al.. The high velocity imbedded layer in the crust of northwestern

China(in Chinese), *Acta Geophysica Sinica* ,**34**,55—63,(1991).

21. Teng Jiwen, Feng Chinfen Li Jinsheng et al. Crustal structure of the central part of north China plain and the HSINTAI earthquake(1)(in Chinese), *Acta Geophysica Sinica*,**18**,255—271,(1974).

22. Teng Jiwen, Wang Guozhen, Liu Daohongl. Crustal structure of the central part of north China plain and the HSINTAI earthquake(2)(in Chinese), *Acta Geophysica Sinica*, **19**,196—207,(1975).

23. Wang Chunyong, Zhamng Xiankang Wu Qingyi et al..Seismic evidence of detachment in north China basin(in Chinese), *Acta Geophysica Sinica* ,**37**, 613—620,(1994).

24. Teng Jiwen,Wei Shiyu, Li Kinshen. Structure of the upper mantle and low velocity layer of the mantle under the HSIGTAI earthquake region on the north China plain(in Chinese), *Acta Geophysica Sinica*, **37**, 613—621, (1994).

25. Feng Rui and Zheng Shuzhen. A comprehensive study on lithosphere structure on north China(in Chinese), *Chinese Sciences Bulletin*, **22**, 1723—1727,(1987).

26. Liu Guodong, Yan Li'en. Draw up and research outline geoscience transect in Eastern China, in: *Committee of Geoscience Transect of State Seismological Bureau,*Beijing: Seismological Press(1994).

27. Teng Jiwen,Sun Kezhong,. The upper mantle structure and LVL of mantle for Qinghai-Xizang plateau region, Annual Report of Laboratory of Dynamical Geodesy(in Chinese), in: *Institute of Geodesy and Geophysics China Academy Science* 110—116, Geological Press(1991).

28. Teng Jiwen, Liu Futian Qian Youli et al.. Seismic tomography of the crust and mantle under the orogenic belts and sedimentary basins of north western China(in Chinese), in: *Advances in Solid Earth Geophysics in China*, 66—88, Beijing: Ocean Press, (1992).

29. Wang Chunyong, Wang Guimei, Lin Zhongyang. A study on fine crustal structure in Xingtai earthquake area based on deep seismic reflection profile(in Chinese), *Acta Geophysica Sinica*, **36,** 445—452, (1975).

30. Wei Siyu, Teng Jiwen, Wang Qianashen et al.. *Lithosphere structure and dynamics for continental margin of Eastern China*(in Chinese), Beijing: Science press,(1990).

31. Liu Guodong. *The structure of earthquake source and dynamical process Nowadays geodynamic and its application*, Beijing: Seismological Press, 70—83,(1994).

32. Zeng Rongsheng, Lu Hanxing, Ding Zhifeng. Seismic reflection and refraction profile crossing Tangshan epicentral region and their implication to seismologenic processes(in Chinese), *Acta Geophysica Sinica* , **31**, 383—398, (1995).

33. Liu Changquan, Tia Shixu. *The velocity structure of Shxi linfen basin, Seismological Research of Shanxi Linfen basin and systematic reduction of hazard*(in Chinese), 231—235, Beijing: seismological publish house,(1994).

34. Teng Jiwen, Yin Zhouxun, Sun Kezhong. *The lithosphere structure and deep internal processes of Xizang plateau in China*, in: Proceeding of International Symposium on the Kakakorum and Kuntun Mountain, 72—88, Beijing: China Meteological Press,(1994).

35. Wenjing Zhao, K.D.Nelson. Deep seismic reflection evidence for continental underthrusting beneath southern Tibet, *Nature*, **366**, 557—559, (1993).

36. Wu Xuanzhi, Wu Chunling et al.. Research on the fine crustal structure of the northern Qitian-Hexi corridor by deep seismic reflection, *Acta Geophysica Sinica* , **3**, Suppl., (1994).

37. Li ng, Zhonghu, An Changqiang Study on 3-D velocity structure and anisotropy beneath the west China from the love wave dispersion(in Chinese), *Acta Geophysica Sinica* ,**34**, 691 707,(1991)

38. Cheng Lihua, Song Zhonghe An Changqiang. Three dimensional shear wave velocity and anistropy of crust and upper mantle in the China north-south earthquake belt(in Chinese), *Acta Geophysica Sinica* , **35**, 374 —385,(1992).

39. Li baiji. Seismic anistropy of the upper mantle in Kunming region, *Advance in Solid Earth Geophysics in China*(in Chinese), 330—338, Beijing: Ocean Press(1992).

40. Hrin.A. Seismic anistropy as an indicator of mantle flow beneath the Himalayas and Tibet, *Nature*, **375**, 571—574, (1995).

41. Zheng Shihua and Gao Yuan. Azimuthal anistropy in lithosphere on the Chinese mainland from observation of SKS at the CDSN(in Chinese), *Acta Seismological Sinica*, **16**, 131—140,(1975).

42. Silver P.G. and Chao W.W.. Shear wave splitting and subcontinental mantle deformation, *J.Geophy.Res.*, **96**, 16429—16454, (1991).

43. Teng Jiwen. *The lithosphere physics and dynamics in Kangdian(Sichuan-Yunnan) tectonic belt*(in Chinese), Beijing: Science Press, (1994).

44. Zheng Ye, Teng Jiwen. The structure of the crust and mantle in the Shuixian-Ma'anshan zone and zone characteristics of the south part of the Tanlu tectonic belt(in Chinese), *Acta Geophysica Sinica*, **32**, 648 —659, (1995).

45. Zheng Ye, Teng Jiwen. Study of the fine structure of the upper mantle in the Shuixian-Qidong area at east part of China(in Chinese), *Acta Geophysica Sinica*, **37**. 553—541,(1994).

46. Teng Jiwen. The lithosphere structure and dynamics of continental margin in southeast China(in Chinese), *Science in China(Series B)*. 866—875 (1994).

Proc. 30th Int'l Geol. Congr., Vol. 20, pp. 41-50
Liu (Ed.)
© VSP 1997

The Propagation of High-frequency Seismic P Waves from Intermediate-Depth Earthquakes in Subduction Zones : Implications for Earthquake Location and Slab Structure

JER-MING CHIU, ZEN-SEN LIAW, YUN-TUNG YANG, AND SHU-CHIOUNG
CHIU, *Center for Earthquake Research and Information , The University of Memphis, Memphis, TN 38152,
U.S.A.*

YU-CHIOUNG TENG
Aldridge Laboratory of Applied Geophysics, Henry Krumb School of Mines, New York, NY 10027, U.S.A.

Abstract

A small amplitude precursor before an impulsive P arrival is consistently observed from intermediate-depth subduction zone earthquakes in local, regional, and teleseismic distances. This small amplitude precursor has been identified as a P-wave train from the particle motion and polarization analysis. Time separations between the two P arrivals can be a few to several seconds which appear to be proportional to the traveling length of the seismic waves inside the subducting slab. A preliminary two-dimensional crustal and upper mantle velocity model including a 60° dipping slab was designed to investigate the relationships between the propagation of high-frequency seismic waves and the internal structure of the subducting slab. Synthetic P-wave seismograms at surface stations of various distances from the trench axis assuming earthquake sources at various depths and different parts of the subducting slab were calculated using a two-dimensional finite-element method. Although it is still very preliminary, results of the modeling suggest that earthquakes at intermediate-depth showing two P arrivals are most likely located in the upper portion and not in the central or lower portions of the subducting slab. The small amplitude first P arrival can be interpreted as a P wave train traveling deeper into the central portion of the slab where seismic wave velocity is higher than that in the upper and lower portions of the slab. The impulsive large amplitude second P arrivals may represent a P wave train traveling along the upper portion of the slab where velocity and thermal gradients are the highest. Basically the second P waves are traveling along a well-developed wave guide near the upper portion of the slab where velocity is slower than that towarding the inner portion of the slab. The multipathing effect becomes apparent only when P waves are traveling beyond certain length inside the slab along up-dip direction such that the arrival times of the two P wave trains become separable.

Keywords: Subduction Zone, Finite-element Method, High-frequency P Waves, Wave-guide

INTRODUCTION

Multiple seismic wave arrivals usually represent multiple paths of seismic wave propagation between an earthquake source and the receiving station. Consequently, important information about structural features between earthquake source and the recording station can be determined if the effects along each path can be identified and studied. Short-period and broadband seismic data from intermediate-depth earthquakes recorded in the North Island of New Zealand have been carefully examined and processed to investigate the effect of the slab structure on the propagation of seismic waves. It has commonly been observed and reported in subduction zone regions that the impulsive P arrival from earthquakes of intermediate-depth is usually proceeded by a small amplitude P arrival (Figure 1), for example, in Louat et al. (1979) and Chiu et al. (1985) for the Vanuatu region, Ansell and Gubbins (1986) for New Zealand, and Suyehiro and Sacks (1979) and Fukao et al. (1983) in Japan region. Sleep (1973) had also reported a 1-3 seconds of arrival time separation between the small amplitude first and the impulsive second arrivals in short-period teleseismic observations. Chiu et al. (1985) reported that the small amplitude first and the impulsive second arrivals can be most clearly seen on seismograms from earthquakes deeper than 70 km, particle motion and polarization analysis concluded that both arrivals are P waves with apparent velocities around 7.7 and 6.9 km/s, respectively (Figure 2). They have proposed that the small amplitude first P arrival is a refracted wave traveling deeper into central portion of the slab where velocity is significantly higher than the upper and lower portions of the slab, and the impulsive second P arrival is a guided wave traveling mainly along the upper portion of the slab where almost constant velocity wave guide is well developed. In this paper, numerical approach using finite-element method is applied to model 2-dimensional wave propagation in a simple slab velocity model that consists of a 60° dipping subduction zone.

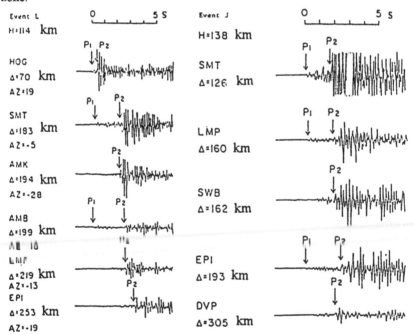

Figure 1. First several seconds of the vertical component seismograms for two intermediate-depth earthquakes recorded by the Vanuatu seismic network showing two distinct P arrivals (P_1 and P_2).

OBSERVATIONS

Previous reports of P-wave precursors (e.g. Louat et al., 1979; Chiu et al., 1985; Suyehiro and Sacks, 1979; Fukao et al., 1983; Sleep, 1973) are mainly based on traditional short-period observations from local and world-wide seismic networks. Figure 3 shows a typical three-component short-period seismograms for an intermediate-depth earthquake recorded recently by the PANDA array (Chiu et al., 1991) on North Island of New Zealand (Chiu et al., 1995). The high frequency nature of the two P waves are apparent in Figure 3. Question is whether the short-period observations can be representative of the wave propagation inside a very complicated subducting slab? Figure 4 shows another example of three-component seismograms for an intermediate-depth earthquake recorded during the same experiment but by a broadband station in New Zealand (Chiu et al., 1995). It is obvious that frequency contents of the two P arrivals are similar in short-period and in broadband observations. The high-frequency nature of the multiple P arrivals may suggest that thin-layered structures must be involved along the ray paths. Thus, frequency content of seismic waves is not considered in the numerical modeling to be presented in the following and the results from the numerical modeling should represent the typical P wave propagation inside the slab.

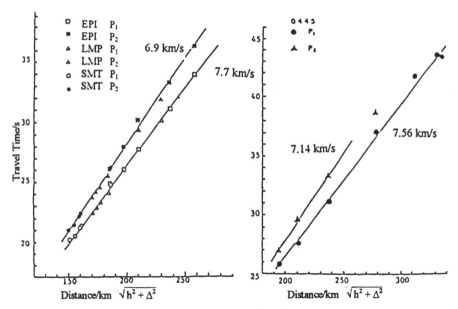

Figure 2. Two plots of travel time versus the straight-line distance between the station and earthquakes showing that the apparent velocity of the small amplitude first arrival is slightly higher than that of the second impulsive arrival, and both arrivals are P waves.

VELOCITY MODEL

The upper 400 km P-wave velocity model of the IASPEI91 (Kennett and Engdahl, 1991) is used to represent the regions outside of a 60° dipping subduction zone. The velocity perturbations inside the subduction zone are described following the patterns of a teleseismic inversion result from Frohlich et al. (1982) and that derived from a thermal model of Sleep (1973). The complicated shallow velocity structure of the slab due to the interactions between the subducting slab and the upper loading plate in this example is neglected simply because of its minor contributions to the generation of

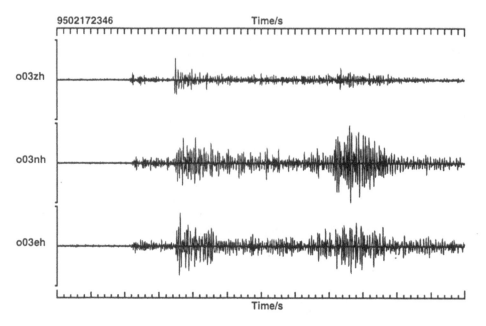

Figure 3. A typical three-component short-period seismogram from an intermediate-depth earthquakes recorded by a PANDA station in New Zealand showing two distinct P arrivals.

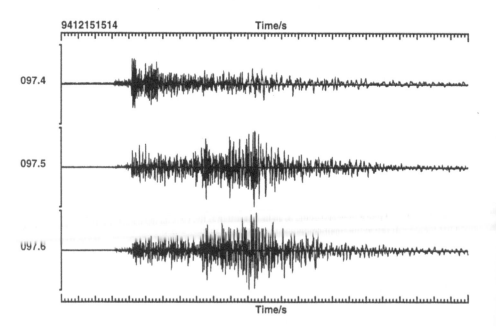

Figure 4. A typical three-component broadband seismogram from an intermediate-depth earthquake recorded in New Zealand showing two distinct P arrivals.

multiple P-wave arrivals. Adjacent layers with similar velocities are merged into a layer of similar velocity to simplify the numerical calculations. Therefore, the final model used in the numerical calculation includes only three layers over a half space (Figure 5).

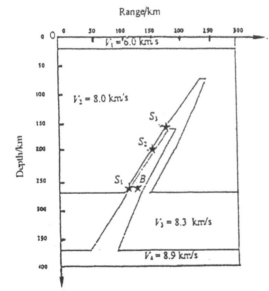

Figure 5. A crustal and upper mantle velocity model derived from the IASPEI91 (Kennet and Engdahl, 1991). Adjacent layers of similar velocities are combined into one layer to simplify the numerical calculations. A 60° dipping subduction zone is also included in the model. Velocity perturbations inside the slab is modified from a thermal model (Sleep, 1973) and from the results of teleseismic inversion (Frohlich et al., 1982).

METHOD : FINITE-ELEMENT MODELING APPROACH

The (x,z) coordinates are set up as shown in Figure 5, with downward positive z-axis representing the depth from surface and positive x-axis toward the right hand-side representing the distances of the observation stations from the reference (0,0) point. The governing equation for an acoustic wave propagating from an applied source function S at the location (x_o, z_o) can be written as

$$\frac{\partial^2 p}{\partial^2 x} + \frac{\partial^2 p}{\partial^2 z} = \frac{1}{c^2}\frac{\partial^2 p}{\partial^2 t} + S\delta(x - x_0)\delta(z - z_0)$$

Or in general,

$$\nabla^2 p - \frac{1}{c^2}\frac{\partial^2 p}{\partial t^2} + S = 0, \quad c^2 = \frac{\lambda + 2\mu}{\rho} \tag{1}$$

where c is the velocity of the compressional wave, μ and λ are the Lame's constants, ρ is the mass density, and S is the applied source function.

In finite element algorithm, the functional formulation corresponding to equation (1) can be written as

$$\chi = \int_v \{\frac{1}{2}[(\frac{\partial p}{\partial x})^2 + (\frac{\partial p}{\partial y})^2 + (\frac{\partial p}{\partial z})^2] - (S - \frac{1}{c^2}\frac{\partial^2 p}{\partial t^2})\}dV \tag{2}$$

After the minimization of the functional (2) with respect to the pressure field, (e.g. Zienkiewicz, 1977), the final system of equation is

$$[M]\{\ddot{p}\} + [K]\{p\} + \{S\} = 0 \tag{3}$$

where

$$M_{ij} = \sum m_{ij}^e \qquad (Pseudo\ mass\ matrix)$$

$$m_{ij}^e = \int_v N_i(\frac{1}{c^2})N_j\,dx\,dy\,dz \tag{4}$$

$$K_{ij} = \sum k_{ij}^e \qquad (Pseudo\ stiffness\ matrix)$$

$$k_{ij}^e = \int_v (\frac{\partial N_i}{\partial x}\frac{\partial N_j}{\partial x} + \frac{\partial N_i}{\partial y}\frac{\partial N_j}{\partial y} + \frac{\partial N_i}{\partial z}\frac{\partial N_j}{\partial z})dx\,dy\,dz \tag{5}$$

N_i, N_j, and N_k are shape functions.

In equation (2), the global pseudo-stiffness matrix is a N×N square matrix, where N is the total number of nodal equations. In order to prevent the stringent requirement of the computer in-core storage of the global stiffness matrix for a large sized model involving enormously long computational time, we adopt a nodal-point-oriented approach (Teng, 1981). The zero matrix elements in the global stiffness matrix are discarded to achieve a fast computational algorithm (Teng, 1989).

In adopting the nodal-point-oriented method and taking advantage of the use of uniform grid of finite elements, this large N×N square matrix for global pseudo-stiffness for a two-dimensional one-degree-of-freedom problem, as the models used in this paper, we reduce the matrix to a row vector with only 9 members disregarding how large the size of the finite element model may be as follows:

$$\{K\} = \{K_1, K_2, K_3, ..., K_9\} \tag{6}$$

For time integration, we use an explicit central difference integration scheme:

$$\{\dot{p}(t + \triangle t)\} = \{\dot{p}(t)\} + \{p(\triangle t)\}\triangle t \tag{7}$$

$$\{\dot{p}(t + \triangle t)\} = \{\dot{p}(t)\} - [M]^{-1}\{p(t + \triangle t)\}\triangle t \tag{8}$$

A finite element model with 300 km in range and 400 km in depth is considered. The size of the element is $x=z=0.15$ km. We specifically address the problem of compressional waves only so that there would be no shear type of Lg waves and their conversions.

Range/km   Source Depth = 150 km

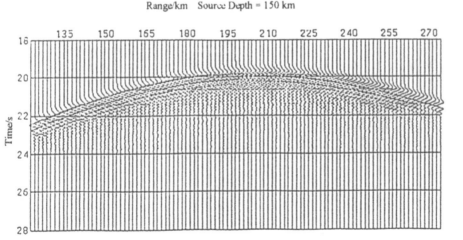

Figure 6. Results of finite-element modeling for a given source located in the upper portion of the slab at depth of 150 km. Precursor is not clear in this example.

Range/km   Source Depth = 200 km

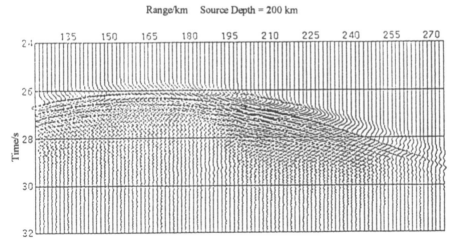

Figure 7. The same as that in Figure 6 with the given source located in the upper portion of the slab at 200 km depth. Precursor starts to become visible at ranges 200 km and beyond

RESULTS

Three intermediate-depth earthquake sources located at depth of 150, 200, and 270 km, respectively, and in the upper portion of the subduction slabs are chosen for the modeling of up-dip propagation of seismic waves.   Synthetic seismograms for compressional wave from each given source and observed at equally spaced surface stations are computed using two-dimensional finite-element method.   Results of synthetic seismograms for the sources at 150, 200, and 270 km are shown in Figures 6, 7, and 8, respectively.   In all three cases, it is clear that precursors of smaller amplitude are visible only at stations at certain distance ranges. For the stations on the left-hand side of these ranges, i.e. further away from the trench axis, P wave takes off

Range/km   Source Depth = 270 km

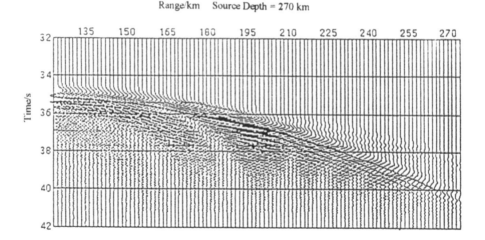

Figure 8. The same as that in Figure 6 with the given source located in the upper portion of the slab at 270 km depth. Well-developed precursors can be clearly seen.

Range/km   Source Depth = 270 km

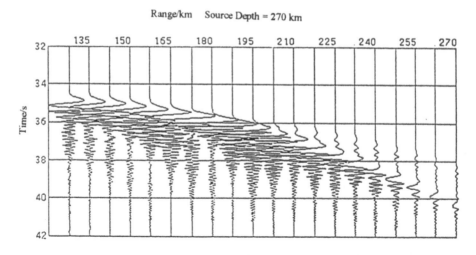

Figure 9. The same as that in Figure 8 with larger interval between stations showing that the critical distance to observe precursor is around 200 km range.

almost immediately from the slab and there are no precursors. For the stations inside these distance ranges, precursors become visible and time separation between the small amplitude precursor and the following impulsive second P arrival is longer for stations further to the right-hand-side, i.e. away from the critical distance as shown in an enlarged display (Figure 9). Since seismic waves are traveling mainly along up-dip direction, it is obvious that time separation between the two P waves is proportional to the length of the traveling path of the P waves inside the subducting slab (Chiu et al., 1985).

Another test was done by shifting the given source horizontally into the central portion of the slab as shown in Figure 5. Results of synthetic seismograms for this example show that there are absolutely no precursors no matter where the station location is (Figure 10). Therefore, the observations of two P arrivals from intermediate-depth earthquakes are closely related to the actual location of the earthquakes inside the slab.

DISCUSSIONS AND CONCLUSIONS

It is obvious that the velocity model we have used in the numerical modeling is probably an over-simplified model. However, major features representing a high velocity slab and its complicated internal velocity structure as shown in Sleep (1973) and Frohlich et al. (1982) have been sustained in our model. Future improvement in velocity model will definitely bring better and more reliable results for the modeling. Nevertheless, the above results shown in Figures 6, 7, 8, 9, and 10 have proved that the internal structure of the subduction zone and the station-source geometry all have their important contribution to the observations of the two P arrivals from intermediate-depth earthquakes. For those earthquakes showing two P arrivals, their hypocenters must be located in the upper portion and not in the inner or central portions of the subduction slab. The small amplitude first P arrival must be a P wave train traveling into the central portion of the slab where velocity is higher than the upper and lower portions of the slab. While the impulsive second P arrival must be a guide wave traveling mainly along the upper portion of the slab. Since significant variations of velocity gradients inside the subducting slab are expected and seismic waves of lower frequencies will most likely traveling transparently across the slab, only high frequency seismic waves, i.e. waves of short wavelength, will be sensitive enough to respond to such scale of velocity variations and thin-layer structures.

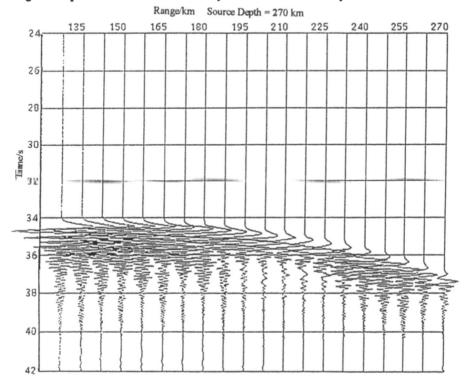

Figure 10. Similar to that in Figures 6 to 9 except that the given source is located in the central portion of the slab at depth 270 km (B in Figure 5). It is clear that no precursor is observed at any station.

ACKNOWLEDGMENT

The New Zealand field project was sponsored by the National Science Foundation under grant EAR-9205235 and partly by the Center of Excellence program at the University of Memphis. This paper is CERI contribution number 313.

REFERENCES

1. Kennett, B.L.N., and E.R. Engdahl, Travel time for global earthquake location and phase identification, *Geophys. J. Int.*, **105**, 429- 465, (1991).
2. Chiu, J.M., Z.S. Liaw, F.T. Liu, Y.T. Yang, E. Roache, M. Reyners, T. Webb, D. Gubbins, and G. Stuart, Seismotectonic studies of the Hikurangi subduction zone in New Zealand, presented in the second International Acoustic Symposium, August, 1995, in Waikiki, Hawaii, (1995).
3. Chiu, J.M., B.L. Isacks, and R.K. Cardwell, Propagation of high-frequency seismic waves inside the subducted lithosphere from intermediate-depth earthquakes recorded in the Vanuatu arc, *J. Gephys. Res.*, **90**(B14), 12741-12754, (1985).
4. Chiu, J.M., G. Steiner, R. Smalley, and A.C. Johnston, PANDA: a simple portable seismic array for local- to regional-scale seismic experiments, *Bull. Seismol. Soc. Am.*, **84**, 1000-101114, (1991).
5. Frohlich, C., S. Billington, E.R. Engdahl, and A. Malahoff, Detection and location of earthquakes in the central Aleutian subduction zone using island and ocean bottom seismograph stations, *J. Geophys. Res.*, **87**, 6853-6864, (1982).
6 Fukao, Y., S. Hori, and M. Ukawa, (1983), A seismological constraint on the depth of basalt-eclogite transition in the subducting oceanic crust, *Nature*, **303**, 413-415, (1983).
7. Louat., B.L. Isacks, and J. Dubois, Anomalous propagation of seismic waves through the zone of shearing contact between the converging plates of the New Hebrides arc, *Nature*, **281**, 293-295, (1979).
8. Sleep, N.H., Teleseismic P-wave transmission through slabs, *Bull. Seismol. Soc. Am.*, **63**, 1349-1373, 1973.
9. Suyehiro, K., and I. Sacks, (1979), P and S wave velocity anomalies associated with the subducting lithosphere determined from travel time residuals in the Japan region, *Bull. Seismol. Soc. Am.*, **69**, 97-115, (1979).
10. Teng, Y.C., Preliminary version of the Aldridge finite-element algorithm for three dimensions, AFEA.3, Project MIDAS Annual Report II, 333-386, (1981).
11. Teng, Y.C. (1989). Three-dimensional finite-element analysis of waves in acoustic media with inclusion, *J. Acoust. Soc. Am.*, **86**, 414-422, (1989).
12. Teng, Y.C., Scattering of transient waves by finite cracks in an anti-plane strain elastic solid, *J. of Compu. Acou.*, **1**, 101-116, (1993).
13. Zienkiewicz, O.C., The finite element method, New York: London, McGraw-Hill Book Co., (1977).

Proc. 30th Int'l Geol. Congr., Vol. 20, pp. 51-61
Liu (Ed.)
© VSP 1997

A Study of Detachment Faults of the Upper Crust in Manzhouli-Suifenhe GGT by the Method of Vertical Seismic Reflection

YANG BAOJUN , LIU CAI, HAN LIGUO, ZHANG HAIJIANG , HE MIN

Department of geophysics, Changchun University of Earth Sciences , Changchun 130026, China

Abstract

Manzhouli-Suifenhe geoscience transect(MS) is one of the eleven transects in China territory determined by global geoscience transect project. Data acquisition, data processing and interpretation of 130 km long vertical seismic reflection were accomplished. The main factor in data acquisition is source energy, and the major steps in data processing are migration and wide-band filtering. The principal purpose in data interpretation is to find out crust structure and its geological meanings. A number of groups of low angle faults, i.e., detachment faults from the upper crust have been found on this profile. These faults have different features but the same formational background, i.e., the compression and torsion stress field in the upper crust in continental margin near the west of Pacific produced by the oblique subduction of Pacific plate in the direction of west. The leading edge of function of the stress field at least reaches near the east of Taikang. Later, the pull-apart activities of these detachment faults arose and formed a series of faulted basins. The detachment faults play an important role in controlling the formation, range, deposit faces, and target estimation on oil and gas of the basin.

Keywords: detachment faults, vertical seismic reflection, interior structure, nappe

INTRODUCTION

Manzhouli-suifenhe geoscience transect is one of the eleven transects in China territory designed by Chinese Lithosphere Committee for carrying out global geoscience transect project(GGT)[1,2,3]. The position of the transect is shown in Figure 1. The purpose of this research is to find out the lithosphere structure of the northeast region of China in East Asia continental margin. It will contribute to the establishment of a geodynamics model of continental lithosphere about west Pacific continental margin[4,5]. In the middle of 1970's, professor J.F. Oliver from Cornell University of U.S.A. organized COCORP (Consortium for Continental Reflection Profiling) with the major objective of studying deep crust structure and mechanism of important geological structures in terms of surface reflection seismic profile, revealing the formation and development of various tectonic units, and finally finding out evolution law of whole crust. Since 1983, France, England ,Germany ,Canada and Australia have successively organized many research groups similar to COCORP and carried out vertical seismic reflection research[6].

Study of Tectonics

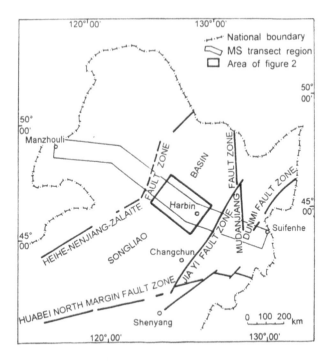

Figure 1. Location of Manzhouli-Suifenhe GGT, China. The Songliao basin is located in the area between the Heihe-Nenjiang-Zalaite fault zone and Mudanjiang fault zone. The 130km long reflection seismic profile is located in the area of Figure 2, its overall direction is perpendicular approximately to the general strike of regional structure of the Songliao basin.

Small receiving distance, short spread, stronger wave energy reflected from the interior of the Earth, abundant processing techniques, and distinct profile are major characteristics of vertical seismic reflection superior to other geophysical methods.

The method of vertical seismic reflection has been used in studying crust reflection patterns including laminated structure, diffraction model, crocodile type, steep and gentle dipping types, bireflections ,fish bone model and weakening model versus depth, etc.[7,8]. Some possible conclusions about reflective lower crust, according to this method, have been proposed as extension structure, invading iron-magnesium, layers filled with fluids, and residual deposit bedding [9,10,11,12]. In Moho research, it is first found that Moho is not just a single interface but a layer with velocity varying progressively or interbeddingly. Besides, it is also discovered that developments of Moho are different in various tectonic backgrounds, and Moho has lateral inhomogeneities[13,14,15]. The application of deep seismic reflection into the study of collision and subduction structure suggests that they appear as crocodile structure in upper crust, which is different from the reflections indicating bright spot in basin and stretching structure[16,17,18]. Seismic reflection was less used in studying strike-slide structure. Seismic reflection features of craton have been compared with those of active zones [19,20]. In brief, studies have shown that the features of seismic reflection show the characteristics of different tectonic units on a large scale[21,22].

Three examples

Recently, vertical seismic reflection has made considerable progress in data acquisition. data processing, and solving geological problems. For example, a seismic reflection profile of 300km long to study deep crust has been successively obtained in Bering Shelf-Chukchi sea using source of air gun array, 25m trace interval, 40-folds, and 18 seconds record. This profile has been used in studying the types and characteristics of continental crust, boundaries of continental blocks ,formation mechanism of deep deposit basin and relationship between orogenic zones and normal faults with low angles,etc.[23]. Another example is the study of INDEPTH (International Deep Profiling of Tibet and the Himalaya)[24]. The deep reflection seismic profile of 300km long was achieved successfully. The data acquisition was combined with wide angle seismic and off-line observation, and also combined with gravity and magnetic data acquisition along the line. The initial interpretation shows that the depth of the top of Indian continental crust subducted beneath Tethyan Himalaya varies between 25-41km. The depth of the Moho can reach about 75km. The rise of Qinghai-Tibet plateau stemmed from the subduction of Indian continental crust. Further examples are found in COCORP and other global deep seismic studies including Project CRATON, Project INDEPTH, Project URSEIS, COCORP superdeep, COCORP in Hyperspace and 3D COCORP[25].

Development trend

Deep seismic reflection is used in the basin's research for the purpose of improving the accuracy. When combined with wide angle seismic sounding, it can effectively control regional structure and velocity distribution. It has been applied successfully in several special problems including the study of crustal normal faults, the nature of earth substances, continental crust structure about transition zone of ocean and continents. It is noticeable that 3D, 3C (three dimensions, three components) technique has made vertical seismic reflection reach an unprecedented accuracy in solving deep geological problems. Besides, people have also paid attention to imaging effects(pre-stack, real 3D) on deep seismic reflection. The progresses made in this aspect, can be referred to the collection of abstracts of "6th International Symposium on Seismic Reflection Probing of the Continents and Their Margins (September, 1994, Budapest, Hungary)" [26,27,28].

Detachment faults

A 130km long profile of vertical seismic reflection (position shown in Fig. 2.) on ME transect was obtained. The study was mainly on several science problems such as the formation mechanism and nature of basement of Songliao basin, controlling effects on other geophysical data in 130km long transect, seismic wave velocity distribution and its geological meanings in the basin, structural features of crust, some basic problems about Moho, oil and gas target estimation of deep layers in Songliao basin, etc. The results demonstrate that there exist many groups of low angle detachment faults in the upper crust of the profile. Although these faults show different features, they have common mechanism that compression and torsion stress field in the upper crust in continent near the west of Pacific is produced by the oblique subduction of Pacific plate in the direction of west. Detachment faults play an important role in controlling Songliao rift basin,

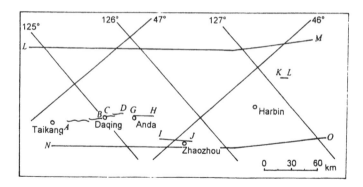

Figure 2. Location of the vertical seismic profile. Vibroseis source was adopted to the *AB*, *CD* segments working along road. Charge source was adopted to the other three segments . 4 station of vibroseis apparatus with 16 times of vibration were used . The correlation length is 14s, 48-folds. 30-folds, 30-folds and 15-folds were used in the *GH*, *IJ* and *KL* segments respectively . The distance between trace of all segments is 50m. The area between broken line *LM* and *NO* belongs to the MS transect region .

notable ones include how the basin was formed, its spatial distribution, deposit faces, and oil and gas target estimation ,etc.

TECHNIQUES OF DATA COLLECTING AND PROCESSING

This 130km long reflection seismic profile consists of the following five segments: segment *AB* and *CD* are 60km long from Taikang to Anda, segment Anda(*GH*), 24km, segment Zhaozhou (*IJ*), 36km, segment Harbin (*KL*), about 10km. During the acquisition, *AB* and *CD* use vibrators, and *GH*, *IJ* and *KL* use dynamites. For vibrators, vehicles not less than 4 are needed to obtain 14-seconds record. Vibration times in fixed spot are also very important . For dynamite, valid energy traveling downward through hole is an important parameter to generate stronger energy. Both the charge amount and the depth of hole are important parameters to collect the data. The interval between composite holes should take into account explosive radius and rock properties. More coverage times don't mean better result because average effects of stack will affect lateral resolution.

The following describes more important steps in data processing. Firstly, selection of valid shots and traces is significant. Secondly, velocities which are used for later ocessing and depth calculation should be reasonably estimated. In this paper, we apply modified Dix formula i.e., calculating layer velocity by the method of weighting and compensation to layer velocity. Thirdly, result of rough stack which supplies ordinary features of major reflection layers can be used to examine processing effects of the methods used and indicates the direction for next works. Fourthly, deconvolution after stack can not be used before the migration since it can degrade the continuity of the events. Fifthly, migration processing is needed when steep interfaces can be seen on raw stack profile. Sixthly, wide band filtering should be applied in order to keep basic features of seismic phases, which are beneficial to the research on dynamic information. Processing result of 130km

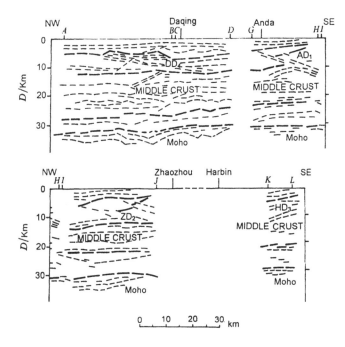

Figure 3. Linear map of the vertical seismic reflection profile. For convenience, the linear map was divided into two parts, i.e. the upper part belonging to *AB*, *CD* and *GH* segments and the lower part to *IJ* and *KL* segments. The two parts have a small segment as common part near the location of *H*, *J*. The thick line indicates the strong amplitude of the reflection seismic phase. The Oblique reflection layers identified by the letters indicate the detachment faults .

reflection seismic data is a time section displayed in wiggle superimposed on variable area. For convenient geological interpretation, a line-shaped profile is shown in Figure3 which is obtained by connecting peak values of seismic phases.

BASIC CHARACTERISTICS OF DETACHMENT FAULTS

For simplicity, the detachment faults on the profile corresponding to Anda, Zhaozhou, Harbin and Daqing are named respectively as AD_1, ZD_2, HD_3 and DD_4. To depict basic features of four groups of detachment faults, we analyzed in detail the faces and reflection layers and obtained some conclusions as shown in Table 1.

The features show that: 1) detachment faults are reflective interfaces to seismic waves. Because they are located in the depth of 4-15km which is equivalent to middle and shallow reflection interface, the seismic records from them have high apparent main frequency and moderate seismic phases. This is beneficial to faces tracing, forward simulating and structure mapping; 2) these four types of detachment faults distribute in the range of 130km. The continuity of seismic faces ,magnitude of apparent dip and

Table 1. Basic characteristics of the detachment faults

detachments characteristics	AD_1	ZD_2	HD_3	DD_4
horizontal length(km)	36	21	25	36
shape	linear	linear	linear	linear
minimum depth(km)	4	6	4	5
maximum depth(km)	15	13	5	13
apparent dip(degree)	27.7	15.6	8.1	12.5
apparent main frequency (Hz)	22	20	23	20
number of phases	2-4	2-3	2	2-4
continuity	good	general	general	intermittent
strength of the amplitude	very strong	general	general	general
interior structure	double layer reflection	intersect	unicline	belt

The vertical seismic reflection profile is different from the linear map obtained from the seismic profile. The first five items in the first column of Table 1 are the geometry characteristics of the detachment faults which were obtained from study of the seismic profile. The next four items are physical characteristics obtained due to the linear map. Especially, the last item indicates the interior structure characteristics obtained in terms of two kinds of data above.

strength of amplitude are all centered at AD_1 and vary in both side ,i.e. continuity becomes poor, apparent dip decreases and strength weakens. These changes indicate that the stress field forming the detachment faults is stronger near Anda; 3) although the basic shapes of these four types of detachment faults are linear, their interior structures are more complicated, ranging from the unicline, the intersect, the double layer reflection, to the belt. This shows that in addition to lateral variations of media containing detachment faults, the effects of stress field forming the faults are complex; 4) it can be seen from Figure 3, that HD_3 may extend to the east, but the depth of convergence interface of the other three groups remains nearly the same, 14km. This depth is consistent with that of shallow earthquake source of Songliao basin in the Northeast of China. Thus, tectonic stability in the region where the detachment faults exist is poor.

GEOLOGICAL INTERPRETATION ABOUT DETACHMENT FAULTS

Northeast of China is mainly affected by two major stages of structure evolution; one is the matchment between early Siberia old plate and Sino-Korea old plate, and the other is the evolution of late littoral Pacific structure region. In the process of matching of Siberia plate and Sino-Korea plate, Songnen micro plate subducted and collided with Daxinganling micro plate along Nenjiang fault, trough was closed, magma in the deep intruded, and a series of reverse faults with NNE or NE direction were formed. Indosinian movement broke up the Palaeozoic tectonic patterns of Northeast of China where the littoral Pacific structure region began to develop [29,30,31,32,33].

Formation and development

The tectonic evolution process of littoral Pacific can be divided into three steps. These are, the period of transition subduction-slip (200-110Ma), the period of oblique extension subduction (110-45Ma) and the period of forward subduction stretch (200-145 Ma) [33,34,35,36,37,38,39].

1) The period of subduction strike-slip. During early and middle Jurassic period (200-145Ma), Japanese Kula plate subducted in NW direction beneath Asia continent. Principal compressional stress within plate changed its direction from SN to NWW-SEE, and generated in upper crust a lot of nappe and strike-slip structures including the detachment faults observed in MS transect region. Because the stress field in this stage was mainly compressional and torsional, the structures it formed expressed as tough shearing deformation and large nappe. Volcanoes were not active[40,41]. During the later Jurassic- early Cretaceous period (145-110Ma), Yizenachi plate in Pacific moved fast here and intersected with epicontinent in small angle and generated oblique subduction, which showed margin transition-shear sliding [42]. In this period, the Songnen micro plate suffered the rift action and produced the Songliao basin which had the characteristics of strike-slip and extension.

2) The period of oblique subduction extension. During early Cretaceous period-Eocene era (110-4.5Ma), Yizenachi plate subducted obliquely to epicontinent of East Asia, which became accretion band from terrain matching. Within the continent sinistral strike-slip was still a major effect, but extension became very remarkable. Songliao basin stretched along the bottom detachment surface under the mechanism of strike-slip and extension. The basin had the nature of settling after rift formation. Under the effect of regional extension stress system ,it formed a group of basins in the form of graben. Tension increased the scale of the structure.

3) The period of forward spread. At the beginning of Eocene era, Pacific plate subducted to Asia plate, the major effects within the plate were spreading and extension. Songliao basin continued to receive sediments, and extension effect was maximum. Different elevation settling movement of Daxinganling and Songliao basin also began from this period.

Application in study of geology

Modern researches about earthquake show that mantle and lower crust of lithosphere can carry out the remote distance transmission of plate boundary forces in the style of plastic flow. Release of the forces mainly appears in multi-earthquake layers in the depth of 10-35km where shear function is principal[43,44,45]. Since Oligocene era, push forces toward north of Indosinian plate became remarkable. Under the effect of remote distance transmission and indirect function of side compression, the upper crust in East Asia would be activated. Besides, many groups of detachment faults within epicontinent of East Asia produced by subduction toward west of Pacific plate would also make the upper crust unstable in this region. Therefore, the depths of these shallow earthquake sources were distributed mainly in 15-20km in this 130km range, with the depth of 15km being common. The oblique subduction of Pacific plate toward west produced the following tectonic units:1) zone of forearc terrain strike-slip and matching. Subduction retreated continuously toward ocean due to the matching of these terrains [46,47]; 2) island arc region. Generally, oblique subduction doesn't generate island arc [48]; 3) back-arc strike-slip structure active region of calcium and alkali magma. At the back of subducted mantle wedge, strike-slip faults usually appeared in the direction parallel to oceanic trench. Along these faults, certain magmaic activities would take place and form some pull-apart basin which has some extension features. This can be called strike-slip-extension composite basin or oblique extension basin[49,50]. The development of oblique subduction of oceanic plate to epicontinent would form nappe fault structures within upper crust at both sides of island arc region and since forearc area is more subducted , this would lead to reverse drawing structure on nappe fault. At back-arc area, reverse drawing faults would not be generated even if nappe faults were quite clear .It is at these different positions of nappe faults in back-arc that effects of stress field have differences in strength . Segment AD_1 is most distinctive on 130km long profile in MS transect region. Basically, it can be inferred that stress field in upper crust generated by oblique subduction of oceanic plate is strongest near Anda in back-arc basin, and the front of effect of stress field at least reaches Taikang in MS region.The pattern and distribution range of detachment faults can be determined by such parameters as lateral length, apparent dip, and buried depth. Together with the magnitude of extension along detachment faults, spacial range of Songliao basin can be estimated. During the period of basin subsidence, a large amount of sediment which has good oil generation condition of Jurassic and Cretaceous deposited in the basin. This shows that detachment faults play an important control role in target estimation of deep oil and gas resources .

CONCLUSIONS

The study of vertical reflection seismic profile of 130km long was accomplished in MS transect . To produce more energy of seismic source, depth is a principal parameter. Effect of increasing coverage times is not always so good . The more effective steps in data processing include selections of valid traces and shots, rough stack, deconvolution, migration and wide-band filtering etc.

Through the study of characteristic parameters of multi-group detachment faults on 130km long reflective seismic profile , we think that as middle-shallow reflection interface, the seismic phases of detachment faults interface have good continuity and high apparent main frequency , which can be easily used in structure mapping. The stress field forming those detachment faults is stronger near Anda. The structure within the detachment faults is complicated. The depth of the convergent interface of bottom is about 14km, and is basically identical to the depth of shallow earthquake source in Songliao basin in the northeast of China.

In early and middle Jurassic period, Kula plate of Japan subducted beneath Asia continent in the direction of NW, which produced a number of structures with the characteristics of nappe, strike-slip in upper crust, including many groups of detachment faults observed in

MS trancect. In early Cretaceous-Eocene era, Yizenachi plate obliquely subducted under Eastern Asia epicontinent ,and Songliao basin extended along the bottom of detachment interface under the mechanism of strike-slip-extension. The shape of detachment faults, distributing range and extension along detachment faults can be linked together to determine the space distribution of Songliao basin. The detachment faults have the control function on target evolution of deep oil and gas in Songliao basin.

From Oligocene era, the force caused by India plate subducting to the north is strong. Because of the indirect function of long-distance transmission effect and side compression, it has active effect on the upper crust of East Asia continent .Many groups of detachment faults of East Asia epicontinent produced by Pacific plate subduction to the west, also have the ability of making the upper crust unstable in the range. The above mentioned make shallow earthquakes concentrate around the depth of 15km.

In different parts of nappe faults produced at back-arc, the strength of stress field is different. In MS transect of 130km long, for the stress field produced in the upper crust by Pacific plate subduction to the west ,the stress field is strongest near Anda in the range of basin of back-arc. The front edge of the stress field at least lies near Taikang of MS transect.

Acknowledgments

Authors thank Prof. L.D Brown for the guidance in the velocity calculation of seismic wave,Prof.K.D.Nelson for the beneficial discussion on the characteristics of nappe structure. Authors give special gratitude to Prof. Teng Jiwen for his guidance in the description of geophysical characteristics , the effect of plate's subduction and comprehensive analysis on seismic data. We thank Prof. Liu Guangding and E. Mensah for the reviews of an early version of the manuscript.

REFERENCES

1. J.W.H.Monger. The global geoscience transects project, *Episode* 9,217-222 (1986).
2. H.J. Gotze and J.W.H. Monger. The global geoscience transects project: achievements and future goal, *Episode* 14,131-138 (1991).
3. K.Fuchs. The international lithosphere program, *Episodes* 13, 239-246 (1990).
4. Wei Siyu, Teng Jiwen, Wang Qianshen, et al. *Lithosphere structure and dynamics in the margin of eastern continent of China*, Beijing, Science Press, in Chinese (1990).
5. Ye Huiwen, Zhang Xingzhou and Zhou Yuwen. Study Lithosphere structure and evolution from the characteristics of ophiolites and blueschists in Manzhouli-suifenhe GGT. In: *Geological study on the structure of Lithosphere and its evolution in M-S GGT (in Chinese)*, M-SGT Geological project group(Eds.). 73-83, Beijing, Seismic Press (1994).
6. Wu Qibin. The recent study of foreign seismic reflection, *Geological Science information* (in Chinese). 12, 85-91 (1993).
7. S.L.Kelmperer, L.D.Brown, J.E.Oliver, et al. Some results of COCORP seismic reflection profiling in the Grenville-age Adirondack Mountains, *J.Earth Sci.* 22,141-153 (1985).
8. R.W.Allmendinger, K.D.Nelson, C.J. Potter, et al. Deep seismic reflection characteristics of the continental crust, *Geology.* 15,304-310 (1987).
9. L.D.Brown. Lower continental crust: Variations mapped by COCORP deep seismic profiling ,*Annal. Geophy.* 5,325-330 (1987).
10. A.G.Green, B.Mikhereit, A. Davidson, et al. Crustal structure of the Grenville front and adjacent terranes, *Geology.* 16,788-792 (1988).

11. S.B.Smithson. Contrasting types of lower crust, *Am Geology. Monogr.* **51**,53-63 (1989).
12. L.D.Brown. A new map of crustal " terranes" in the United states from COCORP deep seismic reflection profiling, *Geophys .J.Int.***105**,3-13 (1991).
13. L.W.Braile and C.S.Chiang. The continental Mohorovicic discontinuity, Results from near-vertical and wide-angle seismic reflection studies. In: *Reflection seismology: A global perspective* . M. Barazangi and L.D.Brown(Eds). Geodyn. Ser.**13**, 257-272. Washington, DC, Am. Geophys. Union (1986).
14. C.M.Jarchew and G.A. Thompson . The nature of the Mohorovicic discontinuity, *Aun. Rev. Earth Planet Sei.* **17**,475-506 (1989).
15. R.Meissner and Th. Wever. The possible role of fluids for the structuring of the continental crust, *Earth Sci. Rev.***32**,19-32 (1992).
16. R.Meissner. Rupture, creep, lamellae and crocodiles happenings in the continental crust ,*Tera nova.* **1**,17-18 (1989).
17. R.Meissner, Th. Wever and P. Sadowiak. Continental collisions and seismic signature , *Geophy. J.Int.* **105**,15-23 (1991).
18. C.Bois. Geological significance of seismic reflections in collision belts, *Geophys. J. Int.* **105**,55-69 (1991).
19. K.D.Nelson. A unified view of craton evolution motivated by recent deep seismic reflection and refraction results, *Geophys. J.Int.* **105**, 25-35 (1991).
20. P.Sadowiak , Th.Wever and R.Meissner. Deep seismic reflectivity patterns in specific tectonic units of western and central Europe, *Geophy. J. Int.* **105**,45-54 (1991).
21. K.D. Nelson, J.A.Arnow, J.H.McBride, et al. New COCORP profiling in the southeastern United States. Part 1: Late Palaeozoic uture and Mesozoic rift basin, *Geology.* **13**,714-718 (1985).
22. R.Meissner and Th. Wever . Nature and development of the crust according to deep reflection data from the German variscides. In: *Reflection seismology: A Global Perspective.* M.Barazangi and L.D.Brown(Eds). Geodyn. Ser.**13**, 31-42. Washington,DC, Am. Geophys. Union (1996).
23. S.L.Klemperer and the Bering-Chukchi Working Group. Preliminary results of deep seismic profiling between Alaska and Russia, Bering Shelf-Chukchi Sea(abs.). In: *6th International Symposium on Seismic Reflection Probing of the Continents and their Margins: Program & Abstracts* . Budapest, Association of Hungarian Geophysicists(Eds.). p.61. Hungary (1994).
24. K.D.Nelson., Zhao Wenjin and Project INDEPTH Team. INDEPTH deep profiling of the Himalayan collision zone(abs.). In: *6th International Symposium on Seismic Reflection Probing of the Continents and their Margins: Program & Abstracts* . Budapest, Association of Hungarian Geophysicists(Eds.). p.41. Hungary (1994).
25. L.D.Brown and Cornell Deep Seismic Research Group. COCORP and global deep seismic profiling(abs.). In: *6th International Symposium on Seismic Reflection Probing of the Continents and their Margins: Program & Abstracts* . Association of Hungarian Geophysicists(Eds.). p.18. Budapest, Hungary (1994).
26. S.R.Taylor and S.M.Melennan. *The continental crust: Its composition and evolution*, Oxford, Blackwell Scientific Publications (1985).
27. S.M.Wickham. Evolution of the lower crust, *Nature* . **333**,119-120 (1988).
28. Budapest,Association of Hungarian Geophysicists(Eds.) *6th International Symposium on Seismic Reflection Probing of the Continents and their Margins: Program & Abstracts.* Hungary (1994).
29. Huang Jiqing, Ren Jishun, Jiang Chunfa, et al. Tectonic outline in China, *Geological Journal* (in Chinese). **2**,117-135 (1977).
30. Li Chunyu and Tang Yaoqing. The division of Asia ancient plate and related problems , *Geological Journal* (in Chinese). **57**, 1-8 (1983).
31. Wang Hongzhen, Yang Sennan and Li Sitian. Basin growth of Mesozoic and Cenozoic era and epicontinent construction development in the east of China and its neibouring region, *Geological Journal* (in Chinese). **3**, 213-223 (1983).
32. Ren Jishun, Chen Tingyu and Niu Baogui. *Construction evaluation and deposition of Continental Lithosphere in the east of China and neibouring region*, Beijing, Science Press, (in Chinese)(1990).
33. Ye Mao, Zhang Shihong and Wu Fuyuan. Paleozoic construction unit and its geological evolution in Manzhouli-Suifenhe GGT, *Journal of Changchun university of earth sciences* (In Chinese). **24**,241-245 (1994).
34. A.Aydin and A. Nur. Evolution of pull-apart basins and their scale independence, *Tectonics.***1**,91-105 (1982).
35. Chen Huanjiang. *The discussion of plate tectonics and the analysis of oil and gas basin*, Shanghai, Tongji University Press, in Chinese (1990).

36. E.V.Artyushkov. Role of crustal stretching on subsidence of the continental crust ,*Tectonophysics*.**215**,187-208(1992).

37. D.H.W. Hutton and R.J.Reavy. Strike-slip tectonics and granite petrogenesis, *Tectonics*. **11**,960-967 (1992).

38. Wang Youlin, Liu Li and Liu Zhaojun. Basement structure of cenozoic basin and construction evolution in Manzhouli-suifehe GGT. In: *Geological study on the structure of Lithosphere and its evolution in M-S GGT, China*. M-SGT Geological project group(Eds.). 26-37, Beijing, Seismic Press, in Chinese (1994).

39. Liu Zhaojun,Wang Youlin, Liu Wanzhu et al. The forming mechanism of Songliao-Hailar basin of Mesozoic era in Manzhouli-Suifehe GGT. In: *Geological study on the structure of Lithosphere and its evolution in M-S GGT, China*. M-SGT Geological project group(Eds.). 14-25, Beijing, Seismic Press, in Chinese(1994).

40. B.P.Werniche. Low-angle normal faults in the Basin and Range province: nappe tectonics in an extending orogen, *Nature*. **291**, 645-648 (1981).

41. G.Radel and H.J.Melosh. A mechanical basis for low-angle normal faulting in the Basin Range,*EOS*.**68**,1149 (1987).

42. J.M.E.Beck. On the mechanism of tectonic transport in zones of oblique subduction, *Tectonophysics*.**93**,1-12 (1983).

43. Huang Peihua and Fu Rongshan. Exploratory in the stage of mantle convection on the bottom of Chinese lithosphere , *Geophysics Journal* (in Chinese). **26**, 39-47 (1983).

44. Wang Shengzu. Multi-layer structure model and plastic current network of Lithosphere in Asia Continent, *Geological Journal* (in Chinese). **67**,1-18 (1993).

45. N.L.Carter and M.C. Tsenn. Flow properties of continental lithosphere, *Tectonophysics*. **136**, 27-63 (1987).

46. S. Shuigu , Shao Jian and Zhang Qinglong. The relation between Nadanhada terrain and Mesozoic Construction of East Asia epocontinent. *Geological Journal*. (in Chinese). **3**, 204-215 (1989).

47. G.Kimura, M.Takahashi and M.Kono. Mesozoic collision-extrusion tectonics in eastern Asia, *Tectonophysics*. **181**,15-23 (1990).

48. L.M.Parfenor and B.A.Natalin. Mosozoic tectonic evolution of northeastern Asia, *Tectonophysics*. **127**,291-304 (1986).

49. G.Kinura. Oblique subduction and collision: forearc tectonics of the Kuril arc, *Geology*. **14**,404-407 (1986).

50. J.P.Platt. *Mechanics of oblique convergence*, IGC. Abs. **29**,413 (1992).

Proc. 30th Int'l Geol. Congr.. Vol. 20. pp. 62-68
Liu (Ed.)
© VSP 1997

Pattern Recognition and Tomographic Seismic Imaging Techniques with Applications to Subsurface Tectonic Structures of the Canadian Shield

ROBERT MEREU, BAISHALI ROY and SONNY WINARDHI

Dept. of Earth Sciences, University of Western Ontario, London, Ontario, Canada, N6A 5B7.

Abstract

Over the past few years a series of long range seismic refraction/wide-angle reflection and near vertical coincident reflection experiments were conducted across the Canadian Shield in eastern Canada. Crustal seismic near vertical reflection experiments have the potential of enabling us to view the subsurface crust in great detail. One of the limiting features of these studies is the poor signal to noise problems which often arise because of scattering and raypath lengthening effects which are caused by small scale lateral heterogeneities within the crust. In our studies, the conventional CDP imaging methods were replaced by much more powerful pattern recognition methods which were applied directly to the CDP gatherers. These methods produced dramatic improvements in the resolution of many of the major subsurface reflectors. Seismic refraction/wide-angle experiments which make use of multiple shots and hundreds of receivers enable us to study the crust from a different perspective. In conducting the tomographic analysis of the data to optimize the subsurface images, new methods of parameterizing the model for lateral heterogeneity were developed which employed triangular block methods coupled with delay time least squares inversion techniques. Greatly improved images were obtained when the regional vertical crustal velocity gradients were subtracted from the conventional velocity solutions to reveal velocity anomaly maps.

Keywords: Pattern Recognition, Tomography, Crust, Moho, Reflection, Refraction, Seismic Image

INTRODUCTION

Over the past ten years a series of long range seismic refraction/wide-angle reflection and near vertical coincident reflection experiments were conducted across the Canadian Shield in eastern Canada. The Central Metasedimentary belt, the Grenville Front Tectonic Zone, the Sudbury Basin and the Midcontinent Rift system were some of the key targets of these studies. In the 1986 Glimpce experiment, air gun sources with a source spacing of 60 meters were used for seismic lines crossing the Great Lakes. In the most recent seismic refraction land experiment, the 1992 Lithoprobe Abitibi-Grenville experiment, 44 explosion shot points were employed at an average shot spacing of 30 km. Each of these shots were recorded by a set of 415 instruments at station spacing of 1 to 1.5 km. The profile length varied from 180 km to 640 km. The dense ray coverage of the area which we achieved with these experiments allowed us to employ seismic tomographic techniques to make a detailed study of the variation of seismic velocities under the tectonic structures of interest. Most of the refraction lines were complemented by near vertical reflection lines using both vibroseis sources on land and air gun sources in lakes. Detailed information on these experiments and methods of analysis were given by Green et al.

(1988), Epili and Mereu (1991), Hamilton and Mereu (1993), and Winardhi and Mereu (1997).

TOOLS FOR ANALYSIS

A series of tools which were designed specifically for seismic refraction/reflection type of data was developed to handle efficiently the huge amount of data resulting from the experiments. These were (i) a printer-plotter program which enabled us to plot large seismic sections on an ink jet printer, (ii) A set of digital filters including polarization filters for use with three-component data and pattern recognition filters for use with CDP data, (iii) an interactive program which enables a user to easily pick first and later arrivals from the refraction data sets on a screen display using both automatic and manual methods and (iv) an interactive ray tracing program for sending any type of ray (P, S, or converted ray) and multiple rays through laterally heterogeneous models.

SIGNAL ENHANCEMENT USING PATTERN RECOGNITION TECHNIQUES

Most near vertical crustal seismic reflection experiments are conducted over regions where in the past the crust has undergone some form of tectonic deformation. The strata are no longer layered but tend to be both laterally and vertically heterogeneous. Seismic energy after travelling through long ray paths in such media arrives at slightly different times at the stations and hence tend to be poorly aligned along their respective travel-time curves. Conventional processing routines which attempt to increase the signal to noise ratio by some stacking technique can in many cases actually destroy the poorly aligned signals such that the resultant subsurface images are not very satisfactory. This occurs even though good reflectors can be seen on some of the shot gatherers and CDP gatherers. Our study (Roy and Mereu 1996) focussed on a signal enhancement technique which is based on a pattern recognition approach, applied to the prestack processing stream in an effort to improve the signal to noise ratio. The discrimination between a signal and a noise is based on attributes such as lateral continutity, amplitude, waveshape, and frequency of the signal. Removal of the noise in the CDP gatherers using this approach reduces the level of noise contamination in the stacked section to a large extent. A slightly different approach for stacking the signals has also been adopted in which the energy of the signals with the same polarity are computed in small moving time windows. This takes into account the misalignment of the signals. An example of the results which we obtained in one of the reflection profiles is shown in Figs. 1a and 1b. Fig. 1a is the image obtained by conventional processing. Fig. 1b shows the improved image using the pattern recognition approach. The image shown is a portion of the dipping shear zone structures from the seismic sections of the Central Metasedimentary Belt of the southeastern Canadian Shield.

TOMOGRAPHIC ANALYSIS

The velocity crustal structure images are obtained from the large seismic refraction/wide-angle reflection data sets using a tomographic analysis method which basically consists of the

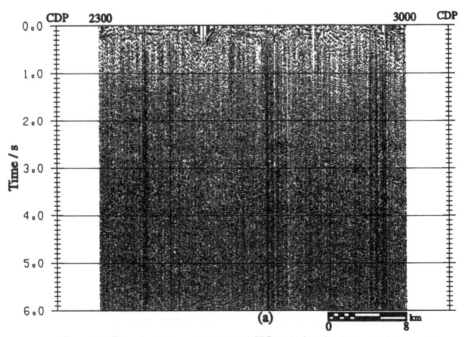

Figure 1a. Example showing a conventional CDP stacked section from a profile across the Central Metasedimentary Belt of the Canadian Shield

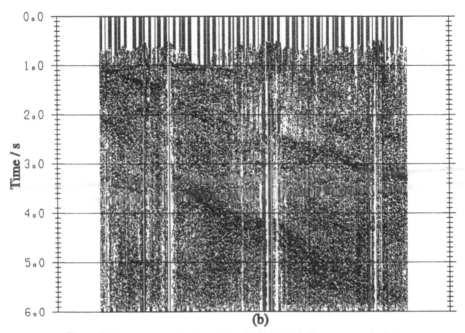

Figure 1b. Example showing the same stacked section as Fig. 1a except this is based on the pattern recognition method of seismic image enhancement

following steps: (i) creating a reasonable starting model, (ii) shooting rays through the model in a forward manner to create a set of theoretical travel-time curves, (iii) carrying out an optimization process which fine-tunes the model until the misfits between the observed and theoretical travel-times are minimized. The initial starting model was obtained by performing a least squares analysis on the complete first arrival Pg data set. This produced an average "standard" crust model which in our case turned out to be a crust with a velocity of 6.13 km/s at the surface increasing linearly to 7.10 km/s at a depth of 40 km. Our observations showed that much of the small scale scatter of points along the travel-time curves were a direct result of lateral and vertical variations in the near surface rocks. We used a delay time method to convert these variations in travel-time data into variations in near surface velocity structure. The remainder of the tomographic analysis for deeper structures was then done using smoothed travel-time curves.

There are a large number of different ways which one can parameterize a model. See for example, Cerveny and Psencik (1984), Zelt and Smith (1992). Each has their advantages and disadvantages. In our studies we chose an updated version of a "triangular block method" (Mereu, 1987) which divides the model into a set of triangular blocks each with its own constant velocity gradient. The triangle was chosen for the model building block because it is very effective in modelling lateral structures and dipping faults. The lateral and vertical velocity variations are expressed as

$$V(x,z) = ax + bz + c$$

$$Grad\ V = a\ i + b\ j$$

where $V(x,z)$ is the seismic velocity at any point in the model and $Grad\ V$ is the constant velocity gradient for each triangle. The constants a,b,c are determined from velocity values which are assigned at each node of every triangle. These three constants and hence velocity gradient are different for each triangle. The constant gradient in the velocity field ensures that all the ray paths traced through each of the triangular blocks are arcs of circles thus enabling one to trace rays through the model very quickly. The ray tracing is achieved by solving ray arc equations with triangular boundary line equations and applying Snell's law at each intersection point. The technique of dividing the model into triangular blocks enables one to work with a relatively small number of parameters. When the structure is simple, the triangles may be large, whereas when the structure is complex, small triangles may be required. The manner in which the model is divided into triangles is not unique. However, numerous tests have shown that in general the travel-times are not very sensitive to this non-unique problem. Other tests which we conducted showed that an optimum number of triangular blocks could be achieved, if it were based on the shot spacing as shown in the example of Fig. 2. Synthetic seismograms are generated from transfer functions which are computed from time delays, free surface effects, geometric spreading effects, as well as all the complex transmission and reflection coefficients that the ray encounters along its path. The computer automatically searches the model for all major travel time branches, numerically codes the branches, determines their end points and then ray traces through the model with appropriate angle ranges for computing arrival-times and amplitudes. The inversion of crustal travel time data is a non-linear inversion problem. It is very easy in any automatic inversion method for models to be created which have unstable solutions or erratic paths. In order to overcome this problem we modified the initial program by making it

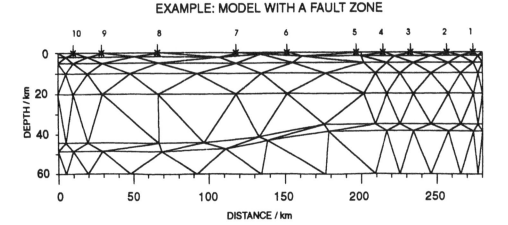

Figure 2. Example showing how the triangular block method may be used to parameterize a velocity model with a fault zone such as that given in Fig. 3a.

interactive.

The new program has features which enable a user to modify triangles and velocities of the model with only a touch of a few keys on the computer keyboard. The inversion is monitored by instant displays of the ray paths and both the theoretical and observed travel-times. During the inversion, a layer stripping or expanding cell approach (White 1989) was used to model the data. In this method the process is started at the top, imaging the shallower crust and then gradually working deeper into the crust to the Moho.

The nature and depth of the Moho is modelled from PmP and Pn observations. In regions where the PmP branch is very short or not observed, the Moho is modelled as thick transition zone. Improved subsurface images were obtained by subtracting an average standard crustal velocity gradient from the velocity section to produce a velocity anomaly section. Techniques for determining the uncertainty of the model were also developed which were based on the relative density of ray coverage and combined with a simulated annealing analysis.

Fig. 3a shows an example of a laterally varying velocity model with a major fault separating a high velocity region from a low velocity region. Fig. 2 shows how this model would be parameterized using triangular blocks. Ten shots were used in this example. Fig. 3b shows how the velocity anomaly image improves the outline of the subsurface fault. Fig. 3c shows a velocity uncertainty section for this example. Here it is seen the most precise velocities are obtained where the ray coverage is most dense. Poor velocity estimates are obtained deep in the crust and at the ends of the model where ray coverage is sparse.

TOMOGRAPHIC ANALYSIS
VELOCITY MODEL

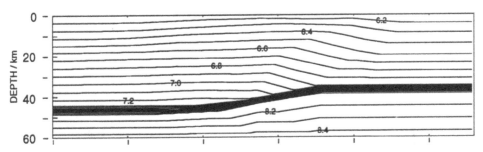

Figure 3a. Example showing a laterally varying velocity model with a major fault separating a high velocity region from a low velocity region.

VELOCITY ANOMALY

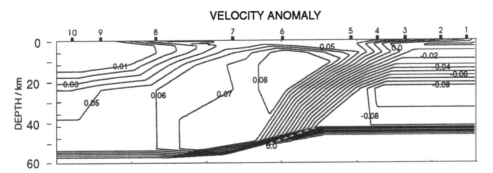

Figure 3b. Example showing a velocity anomaly image. This was obtained by subtracting a standard crust which has a velocity which varies linearly with depth from 6.13 km/s at the surface to 7.10 km/s at the Moho from the velocity model of Fig. 3a.

VELOCITY UNCERTAINTY

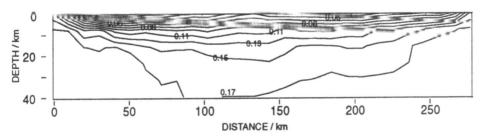

Figure 3c. Example showing how the velocity uncertainty varies with depth throughout the model of Fig. 3a. The uncertainty estimates were based on the density of ray coverage.

Examples of the recent use of our tomographic anomaly method for 4 seismic profiles shot across the Canadian Shield in the 1992 Lithoprobe Abitibi-Grenville seismic refraction experiment are given by Winardhi and Mereu (1997). The most interesting results produced in our analysis was the image obtained for the Grenville Front Tectonic Zone (the boundary between the Superior and Grenville Geological Provinces). This image showed southeast dipping region of anomalous velocity gradients which penetrated the crust right down to the Moho.

CONCLUSIONS

Modern seismic exploration of the crust make use of coincident near vertical reflection and wide-angle/refraction exploration methods. In this paper we have shown two very effective techniques, pattern recognition and tomographic velocity anomaly techniques which can be used to improve the subsurface images of the earth's crust.

Acknowledgements

We thank K. Vasudevan and R. Maier of the LSPF, B. Dunn and J. Brunet from the University of Western Ontario for their technical assistance with regard to the use of the computer hardware and seismic software packages. This study was supported by research grants from Lithoprobe and NSERC Grant A1793.

REFERENCES

1. V. Cerveny and I. Psencik. Gaussian beams in two dimensional laterally varying layered structures. *Royal Astronomical Society, Geophysical Journal*, **78**, 65-91. (1984).
2. A.G. Green, B. Milkereit, A. Davidson, C. Spencer, D.R. Hutchinson, D.R. Cannon, W.F. Lee, M.W. Agena, J.C. Behrendt, W.J. Hinze. Crustal structure of the Grenville Front and adjacent terranes, *Geology*, 16, 688-792. (1988).
3. D. Epili and R.F. Mereu. The Grenville front tectonic zone: Results from the 1986 Great Lakes onshore seismic wide-angle reflection and refraction experiment. *Journal of Geophysical research*, **96**, 16335-16348 (1991).
4. D. Hamilton and R.F. Mereu. 2-D tomographic imaging across the North American mid-continent rift system, *Geophysical Journal International*, 112, 344-358.(1993).
5. R.F. Mereu. An interpretation of the seismic-refraction data recorded along profile SJ-6: Morro Bay-Sierra Nevada, California. In: *Interpretations of the SJ-6 seismic reflection/refraction profile, south-central California, USA, Proceedings of the 1985 CCSS Workshop on Interpretation of Seismic Wave Propagation in Laterally Heterogeneous Terranes, USGS Open File Report 87-73*, A.W. Walter and W.D. Mooney (Eds). 20-37 (1987).
6. B. Roy and R. F. Mereu. Signal enhancement using pattern recognition techniques with application to near vertical crustal seismic reflection experiments, *Geophysical Research Letters*. 23, 1849-1852 (1996).
7. D. White, Two dimensional seismic refraction tomography, *Geophysical Journal International*, **97**, 223-245 (1989).
8. S. Winardhi and R. F. Mereu.Crustal velocity structure of the Superior and Grenville Provinces of the Southeastern Canadian Shield. *Canadian Journal of Earth Sciences*, In press (1997).
9. C. A. Zelt and R. B. Smith. Seismic travel-time inversion for 2-D crustal velocity structure. *Geophysical Journal International*, **108**, 16-34 (1992).

Proc. 30th Int'l Geol. Congr., Vol. 20, pp. 69-82
Liu (Ed.)
© VSP 1997

Two-Dimensional Inversion and Interpretation of Magnetotelluric Data in a Seismic Active Area of Northern Part of North China

ZHAO GUOZE   JIANG ZHAO   LIU GUODONG   TANG JI   LIU TIESHENG
ZHAN YAN
Institute of geology, State Seismological bureau, Beijing 100029, China

Abstract

Two advanced techniques, decomposition of the magnetotelluric (MT) impedance tensor and Rapid Relaxation two-dimensional Inversion (RRI), have been successfully applied to the newly observed MT data. The resultant 2-D electrical structure along Yanggao--Rongcheng profile, which passes through several major tectonic units in northern part of North China, shows that the electrical conductivity structure in the shallow depth agrees well with the regional geological and geophysical information. High conductivity zone is detected discontinuously within middle --lower crust, with obviously lateral inhomogeneity in the crust. Highly conductive zone is also found in the upper mantle. Characteristic distribution of conductivity around the earthquake focal area in the crust is discussed.

Keywords: MT impedance tensor decomposition, 2-D MT inversion, Resistivity distribution, Earthquake

INTRODUCTION

Maganetotelluric measurements were recently conducted at 96 observation sites along 7 profiles in Beijing and its north-west area about 60000 km². The research area is one of the regions with high seismic activities in China. Three major seismic active belts with different strikes conjoin here, namely Zhangjiakou-Bohai sea seismic zone of NW striking, Hebei plain seismic zone of approximately NE trend and Shanxi seismic zone in NNE direction, respectively. According to historical data, there were more than 30 destructive ourthquakes with magnitude larger than 5.0 occurred in this area.

Geologically, the research area belongs to the northern part of North China platform [1], and it can be sub-divided into several secondary tectonic units. Taking Zhangjiakou-Bohai sea seismic zone as a major tectonic boundary in this region, we define Yanshan uplift zone to its north, and Shanxi fault depression, Taihang mountain uplift zone and Cenozoic basin of Hebei plain to its south successively from west to east. The regional tectonic strike is in approximately NE direction. Apart from Hebei plain basin, there are also several small-scale basins within the mountain area.

The results of Magnetotelluric data interpretation presented in this paper come from two MT profiles. Profile I starts from Yanggao, Shanxi province, passing through Shanxi fault depression, Taihang mountain uplift and Hebei plain basin, ends at Rongcheng, Hebei province, which stretches in NW direction. Profile II is located within Yanqing-Huailai basin, trending approximately in E-W direction (Fig. 1). The former one is about 252 km in length, with 19 observation sites along it. Yanggao earthquake on Oct. 19, 1989 with magnitude of 6.1 occurred near it (between site 104 and 105). The latter one has a length of 70 km and 17 observation sites. Shacheng earthquake in 1772 with magnitude of 7.0 occurred near this line.

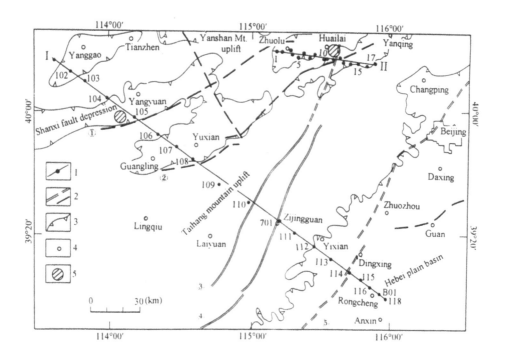

Figure 1   Location of MT profiles and observation sites.   Legend: 1— MT site; 2— fault ((1) Sangganhe river fault; (2) Yuxian-Yanqing fault; (3) Shanghuangqi-Wulonggou fault; (4) Zijingguan fault; (5) Piedmont fault of Taihang mountain); 3— boundary between basin and mountain; 4— city; 5— earthquake epicenter.

Decomposition of the magnetotelluric impedance tensor [2, 3] is used in data analysis and evaluation. It was found that impedance tensor at some sites, processed by the conventional method, was strongly influenced by local telluric distortion. After applying the newly developed decomposition technique, apparent resistivity and phase data which reflect the regional anomaly, the regional strike, and regional skew (which is a dimensionality parameter) are obtained. It is suggested that the regional anomaly is two-dimensional in general, therefore the conductivity structure beneath this area could be simulated by a 2-D model. Then, the decomposed MT data from the two profiles are interpreted by using Rapid Relaxation Inversion (RRI) scheme [4].

And finally, a reliable result of conductivity structure in the crust and upper mantle are obtained. Some features about conductivity structure related to the earthquake focal regions are also discussed.

APPLICATION OF TENSOR IMPEDANCE DECOMPOSITION

Cagniard [5] first put forward the basic theory and the method for estimating MT impedance under the condition of isotropic one-dimensional conductivity distribution. He dealt with MT impedance as a scalar function. To meet the requirements of a two-dimensional or an anisotropic earth model, Swift [6] presented the diagonal minimization method, describing the properties of MT data by parameters such as "strike angle" and "skewness". He suggested to interpret the subsurface structure by using the off-diagonal elements of impedance tensor in case of two-dimensionality. It has been demonstrated, however, from large amounts of MT data, that the apparent resistivity may be seriously distorted (which is called local telluric distortion) when small-scale local conductivity anomalies exist near the earth surface, and the local structures are much smaller compared with the penetration depth [7-11]. In those cases, the conventional method for MT interpretation is inappropriate, even if the regional structure is 2-D.

Magnetotelluric impedance tensor decomposition, with computation scheme proposed by Bahr [2,3,12], Groom & Bailey [13], and graphical study in terms of Mohr circle analysis by Lilley [14], provided new approaches for solving problems mentioned above. In the approach by Bahr, the observed MT data could be presented by a superposition of a regional 1-D, 2-D or 3-D structure and local surface conductivity anomalies [2,3,12]. Taking the regional 2-D as an example, we note that in regional coordinates (x',y'), MT impedance tensor is given by

$$
\begin{aligned}
Z' = A Z_2 &= \begin{bmatrix} a_{11} & a_{12} \\ a_{21} & a_{22} \end{bmatrix} \begin{bmatrix} 0 & Z_{\shortparallel} \\ -Z_{\perp} & 0 \end{bmatrix} \\
&= \begin{bmatrix} -a_{12}Z_{\perp} & a_{11}Z_{\shortparallel} \\ -a_{22}Z_{\perp} & a_{21}Z_{\shortparallel} \end{bmatrix} = \begin{bmatrix} Z'_{xx} & Z'_{xy} \\ Z'_{yx} & Z'_{yy} \end{bmatrix}
\end{aligned}
\tag{1}
$$

where A denotes the distortion matrix, its elements are real and independent of frequency at the low frequency range. Z_2 is the regional impedance tensor, which is not affected by local distortion. It can be seen from equation (1), that the two elements of the measured tensor in each column have the same phase.

In an arbitrary (x,y) coordinate system, assuming that x,y point to north and east respectively, the elements of impedance tensor are the linear combinations of $Z_{\shortparallel}$ and $Z_{\perp}$.

$$Z = R^{\mathrm{T}} A Z_2 R = \begin{bmatrix} Z_{\mathrm{xx}} & Z_{\mathrm{xy}} \\ Z_{\mathrm{yx}} & Z_{\mathrm{yy}} \end{bmatrix} \qquad (2)$$

where R refers to the rotation tensor and R^{T} to its transpose tensor.

By comparing equation (1) with equation (2), one can easily find the condition of determining rotation angle between coordinates (x,y) and (x',y') that is Z_{xx} has the same phase as Z_{yx}, and Z_{xy} has the same phase as Z_{yy}. As soon as the rotation angle is found out, the regional strike angle α_1 is obtained. α_1 is defined as phase-sensitive strike angle, or Bahr's first strike angle α_1 in order to differentiate it from Swift's strike angle α. Similarly, phase-sensitive regional skewness S_2 could also be defined by taking into account of the relation between the information of impedance tensor elements and the rotationally invariant. S_2 is a measure of the three-dimensionality of the regional conductivity distribution, while S_1 refers to the conventional skewness proposed by Swift [6].

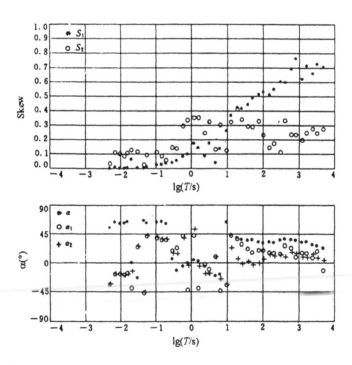

Figure 2 Skewness parameter (top) and strike angle (bottom) for site 106, S_1, α denote Swift's skewness and strike angle, and S_2, α_1, α_2 denote phase-sensitive skewness, strike angle and the modified angle respectively.

In some cases, however, the phases of two elements in one column of the impedance tensor may coincide with each other whereas those in another column may differ very much during the impedance tensor decomposition. Bahr [3], then, improved the technique by adding or reducing the same phase factors in each column, and consequently, yields the modified regional strike angle α_2, namely Bahr's second strike angle.

Decomposition technique is well applied to the observed data taken from northern part of North China, and some important improvement in data quality has been made. Figure 2 gives an example for observation site 106 in profile I. The top figure shows the comparison between conventional Swift's skewness and phase-sensitive skewness. The bottom one shows those for Swift's angle α and phase-sensitive angle α_1, α_2. It is seen that at long period (>10sec), Swift's skewness S_1 has a higher value, maximum 0.7, but phase-sensitive skewness S_2 is much smaller, less than 0.35 in general. That means the data in this site is really affected by local telluric distortion. There is also inconsistency between conventional Swift's angle (α) and phase-sensitive angle (α_1, α_2). At long periods (>10sec), α is about 40 degrees and α_2 varies slightly around 20 degrees. Whereas at short periods (<10s), data scatters, probably reflecting variation of conductivity between the shallow and deep structure.

Figure 3 shows the apparent resistivity and phase data before and after decomposition, again for site 106. It is noticed that before decomposition (left), the apparent resistivity drops abnormally at long period, and the phase data reaches almost to 90 degrees, but they look obviously more reasonable after decomposition (right).

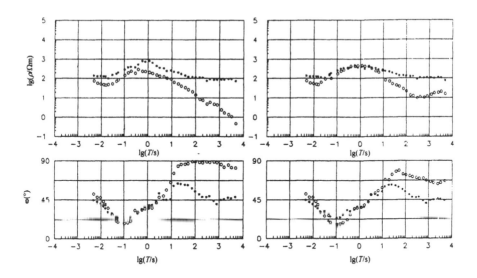

Figure 3 Apparent resistivity (top) and phase data (bottom) for site 106, in the rotated angle by Swift method (left) and rotated by 20 degrees (right), where " *" and " o" denote polarization in *xy* and *yx* direction respectively.

Figure 4 gives stereograms of skewness obtained by the two different techniques, proving the effect of decomposition along the whole profile instead of a single site. It is clear that Swift's skewness S_1 (top) varies abruptly along the profile with higher

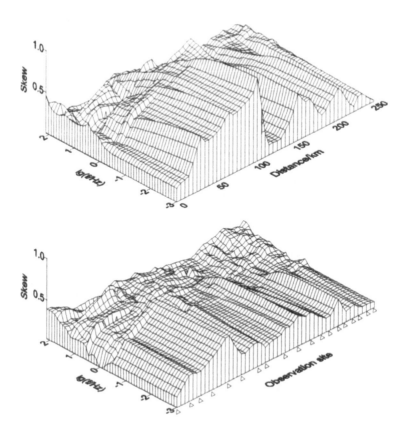

Figure 4 Stereograms of skewness along profile I. Top: Swift's skewness, Bottom: phase-sensitive skewness

values, some reach up to 1.0, while phase-sensitive skewness S_2 (bottom) varies much smoothly, with values mostly below 0.3. It is illustrated that after decomposition, the regional conductivity anomaly along profile I could be modeled by a 2-D structure.

TWO-DIMENSIONAL INVERSION OF MT DATA

A variety of MT 2-D inversion methods have been proposed [15-20]. Among those, Rapid Relaxation Inversion (RRI) algorithm, developed recently by Smith and Booker[4], has its superiority. It runs much fast, uses substantially less memory, and can be executed on an ordinary PC computer such as 386. The efficiency of RRI method is due to that the conductivity perturbation beneath each observation site is calculated by a similar technique to 1-D inversion. The lateral gradients of the electric and magnetic fields in 2-D wave equations are approximated by the values obtained from previous iteration, so only one forward problem needs to be solved to calculate the data residuals and the electric and magnetic fields inside the model for each iteration. The computer memory is, therefore, mainly used to store the model parameters and the field values at each node. It is of great advantage for calculating some complex earth models with more observational data.

We applied RRI method to the inversion and interpretation of the two MT profiles mentioned previously. The input data are the decomposed data, i. e., resistivities and phases in the principal axis for both TE and TM mode. Static shift may occur in apparent resistivity curves at some MT sites, and it is effectively corrected during the inversion by adjusting the "distortion parameters". The general procedure for data interpretation is as follows: firstly, MT data with two different modes (TE&TM) is determined in combination with analysis of geomagnetic induction vectors and some other data. At site 106, for example, the regional strike points to SEE, as it is known from the regional distribution of geomagnetic induction vectors and geological information which is in keeping with the calculated rotation angle, so that we define apparent resistivity and phase data in xy direction as TM mode, and those in yx direction as TE mode. Secondly, the apparent resistivity and phase data for TE and TM mode are separately inversed. As a result, initial conductivity distribution along the profile is obtained, and some " control parameters" are carefully handled which is important for leading the iterations to converge stably. After that, the joint inversion for both TE and TM mode is performed. Above steps are usually repeated several times until fairly good fitting and a reasonable final 2-D model are achieved.

Figure 5 shows the pseudosections of observed apparent resistivity and phase data along profile I for both TE and TM mode. Figure 6 shows the pseudosections of calculated responses after 2-D joint inversion. In each diagram, the ordinate is logarithm of period in second, and the abscissa is the distance along the profile in km. Numbers on the top boundary denote the site No., numbers inside the top panels are the values of apparent resistivity (in Ω m), and those inside the bottom ones are the values of phase data (in degree). It can be seen that there is fine consistency between observed data and model responses, which ensures us a sufficient confidence of the final model. Figure 7 shows the final 2-D conductivity model achieved by RRI joint inversion, where the ordinate is depth in km, the abscissa is again the distance along the profile in km. Numbers on the top boundary have the same meaning as in Fig.5. And those inside are logarithms of model resistivity (in Ω m).

ELECTRICAL CONDUCTIVITY STRUCTURE OF THE CRUST AND UPPER MANTLE

The interpretation of the two-dimensional model along profile I provides us some new information and understanding about the electrical conductivity structure of the crust and upper mantle, which are described in the following.

1) There exists a surface conductive layer at some observation sites with different thickness, where Cenozoic sediments are well developed. For instance, the surface conductive layer appeared near site 102 corresponds to Yanggao basin. Similarly, that between site 104 and 105, that near site 107 and 108, and that to the east of site 115 correspond respectively to Yangyuan basin, Yuxian basin and Hebei plain basin. An abnormally highly resistive zone (with resistivity over 1000 Ω m) is noted near site 109 and 701, which could be explained by Zijingguan magma zone. Furthermore,

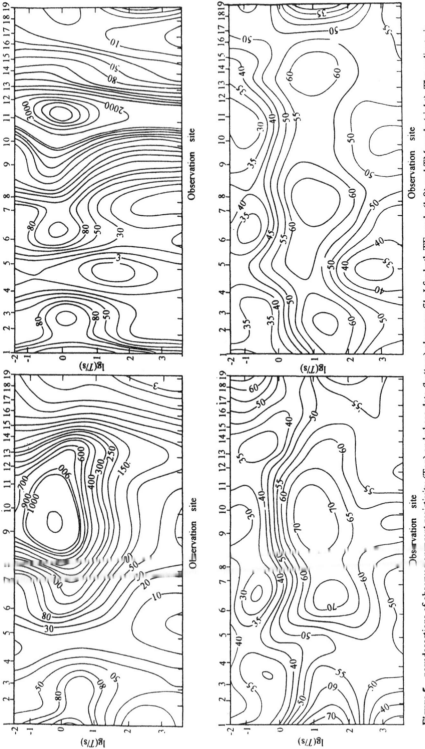

Figure 5 pseudosections of observed apparent resistivity (Top) and phase data (bottom) along profile I for both TE mode (left) and TM mode (right). The ordinate is logarithm of period in second, and the abscissa is the distance along the profile in km. Numbers on the top boundary denote the site number. Numbers inside top panels are the values of apparent resistivity in Ω m), and those inside bottom ones are the values of phase data (in degree). The meaning also suits next figure.

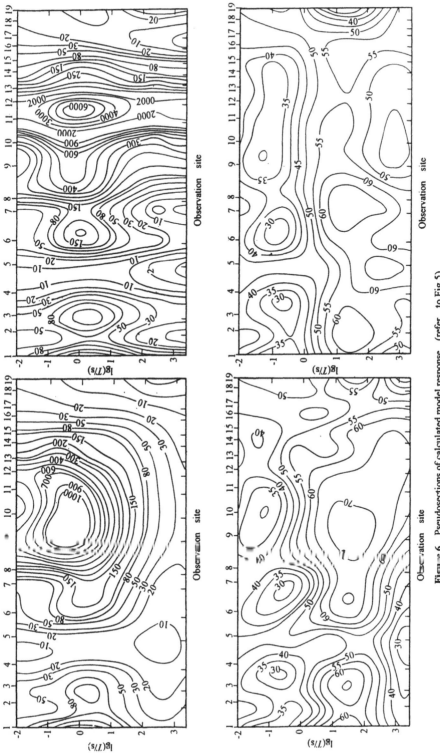

Figure 6 Pseudosections of calculated model response (refer to Fig.5)

there is also a relatively resistive zone near the surface at some sites within Hebei
plain basin, such as site 118, which is situated at Rongcheng uplift.

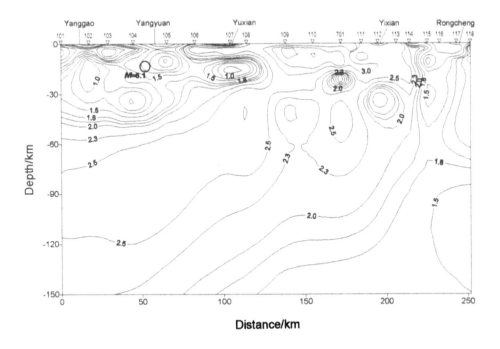

Figure 7 Final 2-D conductivity model along profile I by joint RRI inversion. Number inside the panel is
representative of logarithm of resistivity (in Ω m). Note the focus of Yanggao earthquake (M=6.1, Oct.19, 1989)

2) A high conductivity zone is detected within the middle or lower crust with a
general tendency of decreasing in depth from northwest to southeast. This feature is
consistent with that of Moho discontinuity in this region defined seismically whose
depth varies southeastwards from about 40 km to about 35 km. The MT profile,
according to its different characteristics, can be divided into three segments. The
crustal high conductivity zone is well developed within the first segment (to the
northwest of site 108) with a complex inner structure. While it is less predominant in
the middle segment (from site 109 to 114). Much of the crust in the last segment (to
the southeast of site 115) appears not as resistive as that in other segments, the high
conductivity zone is buried at a shallower depth, and extends upward to the upper
crust.

3) A northwest-dipping highly conductive zone is also revealed in the upper mantle.
It lies at a depth of less than 100 km beneath Hebei plain basin, the southeast end of
the profile, and deepens remarkably to 150 km or more beneath Shanxi fault
depression, the northwest end of the profile, probably correlating to the upper mantle
asthenosphere.

4) Several distinct electric boundaries can be noted in the cross section, such as near
site 105—106, site 108—109, site 701—111, where Sangganhe fault zone,
Yuxian—Yanqing fault and Zijingguan fault pass through respectively according to

relevant geological information. The largest resistivity contrast is found between site 114 and 115, coinciding with the large-scale piedmont fault zone of Taihang mountain. There is an exception, however, no resistivity contrast appears near Shanghuangqi—Wulonggou fault, as known from geological data. The reason is probably that on both sides of the fault there are similar highly resistive bodies to which MT method is not sensitive.

5) A transition zone with a steep rising in depth for the upper mantle highly conductive zone appears beneath the section from sites 701 to 114, which is coincident with the well-known high gradient zone of the tremendous gravity anomaly running through East China. It is proved again that the lateral inhomogeneity of upper mantle structure could be an important factor for the regional gravity anomaly[21].

There is a good agreement between MT result and other geophysical data. For example, according to the velocity distribution obtained from deep seismic sounding data (communicated with Zhu Zhiping) low velocity zones are found in both middle and lower crust to the west of Sangganhe fault, but they are absent between Yuxian-Yanqing fault and Zijingguan fault where the crustal high conductivity zone also disappears. The low velocity zone is buried at a greater depth between Zijingguan fault and Taihang mountain piedmont fault zone, while it appears in the upper crust to the east of Taihang piedmont fault. Those features are consistent with that of MT result. As for geothermal data gathered in this region, though they are inadequately and unequally distributed, there is still a general trend of higher heat flow values in Yuxian basin of about 60 mw/m^2, which is higher than the average value (communicated with Wu Qianfan).

CONDUCTIVITY DISTRIBUTION NEAR THE SEISMIC FOCI AREAS

As it is known, Yanggao M=6.1 earthquake (Oct. 19, 1989) and Shacheng M=7.0 earthquake (1720) occurred near MT profile I and profile II respectively. It is indicated from Figure 7 that the crustal high conductivity zone developed very well near the earthquake foci area and it consists of several interconnected small high conductivity zones. The earthquake with its focal depth of 14 km occurred near the boundary of pronounced resistivity diversity beneath site 104 and 105, where a high resistivity body exists on its northwestern side and a low resistivity body on it southeastern side. As it is known from geological data, however, the basement here on both sides is composed of the same rocks of Archaean Sanggan group. So it can be reasonably considered that the specific conductivity distribution in this area is related to the seismicity.

Figure 8 shows the cross section along MT profile II where Shacheng earthquake occurred. Obviously there is a thin conductive layer near the surface which is thought to be the sedimentary fill in Yanqing-Huailai basin. The upper crust is resistive, apart from a local area near site 309, where low resistivity appears, and coincides with the relatively high heat flow value (about 60 mw/m2). The most interesting fact to be noted is the large resistivity contrast belt in the cross section. It appears in upper crust at the west end of the profile and dips gradually to the east . The most likely explanation is detachment in the upper crust, beneath which there exists an abnormally high conductivity body. The detachment fault, however, seems to be interrupted beneath site 161, where the resistivity contrast belt appears steeply. It is inferred that a deep fault exists with high-angle inclination. (It is also possible that Zijingguan fault and Yuxian-Yaqing fault extend deeply to the crust and intersect here). Shacheng earthquake occurred just near the conjunct area of the different faults and the corner of high conductivity body.

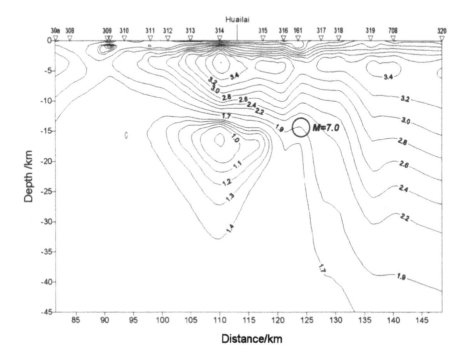

Figure 8 Conductivity model along profile II within Yanqing-Huailai basin, near Shacheng (*M*=7.0) earthquake foci

CONCLUSION

Magnetotelluric data along two profiles, Yangao-Rongcheng profile (I) and the E-W trending profile (II) within Yanqing-Huailai basin, are analyzed by using MT impedance tensor decomposition technique. It is illustrated that the regional conductivity anomaly along the two profiles can be simulated by a two-dimensional model. The strike angle for regional structure are estimated and the apparent resistivity and impedance phase data in two polarization directions are determined. A rapid inversion method (RRI) are used in joint interpretation for both apparent resistivity and phase data and for both TE and TM modes. The sub-surface conductivity models with a high confidence limit are finally obtained.

MT result along profile I shows that the high conductivity zone discontinuously developed in the middle or lower crust with a distinct lateral variation. The upper mantle highly conductive zone changes abruptly with its depth from less than 100 km (beneath the Cenozoic basin of Hebei plain) to more than 150 km (beneath Taihang mountain area). The transition zone is coincident with the high gradient of the tremendous gravity anomaly in East China.

Special attention is drawn to the local conductivity structure where Yanggao earthquake and Shacheng earthquake occurred. It seems that near both earthquake focal areas there exist highly conductive zone in association with sharp boundaries of resistivity contrast. The image of 2-D conductivity distribution offers us the important information about the deep structure of earthquake foci area, which is certainly of great help for our understanding of the earthquake development and occurrence. It should also be pointed out, however, that the genesis of the highly conductive zone may differ in different tectonic environment and at different depths, and some comprehensive study needs to be made in the future .

Acknowledgements

This study is jointly supported by the National Scientific and Technological Committee and the State Seismological Bureau of China under the grants NSTC 85-907--02-03 and SSB 85--02-03. It is also partly supported by the National Foundation of Sciences. The authors wish to acknowledge Prof. J. Booker and Dr. K. Bahr for the helpful discussion on data analysis and for providing their computer codes. We also greatly appreciate the suggestion and encouragement of Prof. Liu Guangding for preparing the manuscript. Thanks are given to other members of the MT group for the field work.

REFERENCE

1. Ma Xingyuan, *Lithospheric dynamics ATLAS of China*, Beijing, Cartographic Publishing House (1989).
2. Bahr K., Interpretation of the magnetotelluric impedance tensor: regional induction and local telluric distortion, *J. Geophysics* **62**, 119—127 (1988).
3. Bahr K. Geological noise in magnetotelluric data: a classification of distortion types, *Phys. Earth Planet Inter.* **66**, 24—28 (1991).
4. Smith, J. T., Booker, J. R., Rapid inversion of two-and three-dimensional magnetotelluric data, *J. Geophys. Res.*, **96**, B3, 3905 3922 (1991).
5. Cagniard, L., Basic theory of the magnetotelluric method of geophysical prospecting, *Geophysics*, **18**, 605—635 (1953).
6. Swift C. M., A magnetotelluric investigation of an electrical conductivity anomaly in the south-western United States, Ph. D. Thesis M.I.T. Cambridge, MA (1967).
7. Jones A. G., Static shift of magnetotelluric data and its removal in a sedimentary basin environment, *Geophysics*, **53**, 967—978 (1988).
8. Spitz, S., The magnetotelluric impedance tensor properties with respect to rotation, *Geophysics*, **50**, 1610—1617 (1985).
9. Zhao, Guoze, Zhan, Yan, Interpretation on the MT data in tectonically complex regions, in: *Investigation and application of contemporary geodynamics*, Beijing, Seismological Press (1994).
10. Berdichevsky, M. N., and Dmitriev, V.I., Distortion of magnetic and electric fields by near-surface lateral inhomogeneities: *Acta Geod. Geoph. Mont. Hung.*, **11**, 447—483 (1976).

11. Jones A. G., Groom R. W, and Kurtz R. D., Decomposition and modelling of the BC87 dataset, *J. Geomag. Geoelectr.*, **49**, 1127—1150 (1993).

12. Eisel M., K. Bahr, Electrical anisotropy in the lower crust of British Columbia: an interpretation of a magnetotelluric profile after tensor decomposition, *J. Geomag. Geoelectr.*, **45**, 1115—1126 (1993).

13. Groom, R. W, Bailey, B. c., Decomposition of magnetotelluric impedance tensor in the presence of local three-dimensional galvanic distortion, *J .Geophys Res*, **94**, 1913—1925 (1989).

14. Lilley, F. E. M. Three-dimensionality of BC87 Magnetotelluric Data Set Studies using Mohr Circles, *J. Geomag. Geoelectr.* , **45**, 1107—1113 (1993).

15. Weidelt, P., Inversion of two-dimensional conductivity structure, *Phys. Earth. planet. Inter*, **10**, 282—291 (1975).

16 Jupp. D. L. B., Vozoff. K., Two-dimensional magnetotelluric inversion, *Geophys, J. R. astr. Soc.*, **50**, 333—352 (1977).

17 Ku, Chao. C., Numerical inverse magnetotelluric problems, *Geophysics*, **41**, 276—286 (1976).

18. Sun Bijun, Chen Leshou, Wang Guange, et al, Inversion of magnetotelluric data, *Acta Geophysica Sinica*, **28**, 218—229 (1985).

19. Zhao Guoze, Liu Guodong, A new inversion scheme for two-dimensional magnetotelluric modelling, *J. Geomag. Geoelectr.*, **42**, 1209—1220 (1990).

20. Wu, N., Booker, J. R., Smith, J. T., Rapid two-dimensional inversion of COPROD2 data, *J. Geomag. Geoelectr*, **45**, 1073—1087 (1993).

21. Zhao Guoze, Zhao Yonggui, The formation mechanism of the North China Plain Basin: a comprehensive analysis of the thermal and gravitational action, *Acta Geologica Sinica*, **60**, 102—113 (1986).

Proc. 30th Int'l Geol. Congr., Vol. 20, pp. 83-94
Liu (Ed.)
© VSP 1997

A Study of Geoelectrical Structure of Crust and Upper Mantle in the Chinese Mainland

LI LI

Institute of Geophysical and Geochemical Exploration, Ministry of Geology and Mineral Resources, Langfang 065000, China.

Abstract

According to the results of magnetotelluric sounding the geoelectrical structure of crust and upper mantle was studied. The variation of resistivity at depth 30 km seems irregular and proved that the geoelectrical structure of upper crust is complex, it consists of blocks of various size and different resistivity. Generally, most of low resistivity areas correspond to high geothermal areas. The geoelectrical structure at depth of 90 km is simpler, some low resistivity anomalies are distributed against high resistivity background, of them the Songliao, the eastern part of North China Platform and western part of Yangtze Platform are connected to each other, formed an immense northeast—southwest trending low resistivity zone. Daxing-ganling—Taihanshan gravity gradient zone, Fen—Wei graben, Longmenshan—Panxi tectonic zone are located at the western fringe of the low resistivity zone; Yilan—Yitong fracture zone and TanLu fracture zone are situated on the eastern fringe, it was inferred that the low resistivity zone was related to the formation of tectonic zones abovementioned. Low resistivity zones at 90 km depth are distributed at the thin down regions of lithosphere, where partial melting mantle material may exist. The low resistivity is taken as background at depth of 150 km, on which scattered blocks with high resistivity are distributed, the axial line of high resistivity blocks formed some "tectonic line" and cut apart Chinese mainland into a few geotectonic units. The low resistivity layer in crust at shallow depth exists in most active tectonic areas. The focuses of shallow—focus earthquakes are generally located over low resistivity layer in crust. The depth of low resistivity layer in upper mantle varies greatly, from shallowest 50—60 km to deepest more than 200 km, the average depth in eastern regions is about 100 km, and that in western is nearly 120 km. A great northeast—southwest trending upwarping zone of low resistivity layer in upper mantle was discovered, from Songliao of Northeast China through east fringe of North China Platform to southwest along Fen—Wei graben, in the final to the west fringe of Yangtze Platform. Upwarping region of low resistivity layer in upper mantle mostly corresponds with high heat flow region and strong earthquake region.

Keywords: Magnetotelluric sounding, Crust, Upper mantle, Resistivity, Chinese mainland

INTRODUCTION

This study's data were based on the recent ten years' results of deep magnetotelluric sounding of Ministry of Geology and Mineral Resources, State Seismological Bureau, Chinese Academy of Science and related colleges and institutes[1-24]. About 1000 deep stations were collected, gethered and processed to yield resistivity values at 30 km, 90 km, 150 km depth, the depth values of low resistivity layers in the crust and in the upper mantle. Furthermore, the data are averaged within 1°×1° of longitude and latitude grid. Since the distribution of deep magnetotelluric sounding stations was uneven, , with yet some blank areas where no station of deep magnetotelluric sounding exists, and at some areas the low resistivity layers in the crust and in upper mantle are estimated by heat flow values[25], so we will only provide a general picture of deep geoelectrical structure in the Chinese mainland.

GEOELECTRICAL STRUCTURE AT DIFFERENT DEPTHS

The resistivities at 30 km, 90 km and 150 km were used in order to research the geoelectrical structure in the crust and upper mantle. They represent the electrical property within the crust, at the bottom of lithosphere and under lithosphere.

At 30 km depth.
The variation of resistivity at 30 km depth seems irregular and proved that the electrical structure of crust is complex, it consists of blocks of various size and different resistivity. According to the results of experiment under high temperature and pressure in rock resistivity, the low resistivity anomalies were mostly due to strata containing fluid or graphitization[26, 27]. Most of low resistivity anomalous zones correspond to high geothermal areas.

At 90 km depth.
The geoelectrical structure at depth of 90 km is simpler, low resistivity anomalous zones are distributed against a high resistivity background (Figure 1), they are Songliao block, eastern part of North China Platform, westrn part of Yangtze Pltform and its east fringe, eastern part of South China Fold system, Tengchong Folding zone, Qing—Zang Plateau, Qilian Fold system, Yinchuan graben in the western fringe of Ordos, Tianshan—Beishan Folding zone, Altay Folding zone and Banggong Co—Gar area. Of them, the Songliao block, the eastern North China Platform and western Yangtze Platform are connected to each other, form an immense northeast—southwest trending low resistivity zone. Daxinggenling—Taihangshan gravity gradient zone, Fen—Wei graben, Longmenshan—Panxi tectonic zone are located to the western fringe of this low resistivity zone, and Yilan—Yitong Fracture zone and Tan—Lu Fracture zone situated on the eastern fringe of this low resistivity zone. It was inferred that the low resistivity zone was related to the formation of tectonic zone abovementioned. This low resistivity zone coincided with the Songliao—Yangtze positive anomalous zone of the satellite magnetic map composed by An Zhenchang[28]. Inferring from that, the anomalous zone was an upwarping zone of upper mantle. The low resistivity zone of 90 km depth is distributed mostly in the thin—down regions of lithosphere, where partially melted mantle materi-

al may exist.

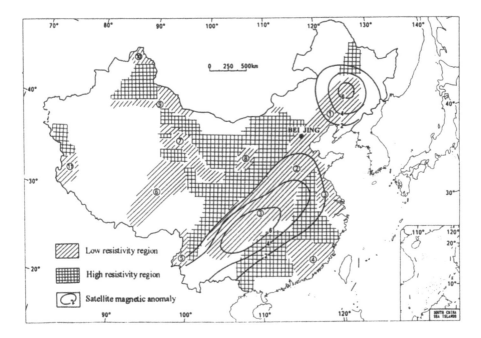

Figure 1. Geoelectrical structure at 90 km depth in the Chinese mainland, showing low resistivity anomalies: ①. Songliao block, ②. eastern part of North China Platform, ③. western part of Yangtze Platform and its east fringe, ④. eastern South China Fold system, ⑤. Tengchong Folding zone, ⑥. Qing—Zang Plateau, ⑦. Qilian Fold system, ⑧. Yinchuan graben in the western fringe of Ordos, ⑨. Tiangshan—Beishan Folding zone, ⑩. Altay Folding zone, ⑪. Banggong Co—Gar area. Isolines show the Songliao—Yangtze positive magnetic anomaly[20].

At 150 km depth.

The low resistivity is taken as the background at 150 km depth, on which scattered blocks with high resistivity are distributed (Figure 2). At this depth, the mantle materials in the majority of areas are in the partial melting state. The high resistivity regions were locally shown only in the northern Songliao block, eastern Jilu nuclear area, the conjunction between Yangtze Platform and South China Fold system, southern Youjiang Folding zone, Longmenshan—Panxi tectonic zone, Qaidam block, Qilian Folding zone, Altay—Tianshan area and west Kunlunshan. It is inferred that the high resistivity areas abovementioned are "hard blocks" consisting of remainning infusible mantle materials.

These high resistivity blocks possibly are the marking of deep tectonic zone, if connect together the axial line of high resistivity blocks, then we get six interesting " tectonic lines" $F_1 - F_6$. Where F_1 started from west Kunlunshan, passing through Qaidam, Qilian to east of Jilu nucleus, formed a west—east trending " tectonic line", cut apart Chinese mainland into south and north part. Jung-

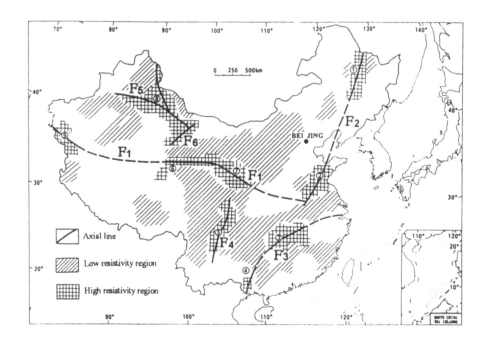

Figure 2. Geoelectrical struture at 150 km depth in the Chinese mainland. Some high resistivity blocks are distributed against a low resistivity background, showing high resistivity blocks: ①. northern Songliao block, ②. eastern Jilu nuclear area, ③. the conjunction between Yangtze Platform and South China Fold system, ④. southern Youjiang Folding zone, ⑤. Longmenshan—Panxi tectonic zone, ⑥. Qaidam block, ⑦. Qilian Folding zone, ⑧. Altay—Tianshan area, ⑨. west Kunlunshan. The axial lines of high resistivity blocks formed some" tectonic lines": F_1-F_6.

gar. Tarim. North China Platform are located at its north side and Qing—Zang Plateau. Yangtze Platform. South China Fold system are located at its south side. The axial line of northern of Songliao block southward connected with Tan   Lu Fracture zone formed a northeast—southwest trending "tectonic line" F_2. The axial line of high resistivity block between Yangtze Platform and South China Fold system is extended southwest'ward and connected with You-jian Folding zone to form "tectonic line" F_3. it is possibly the deep boundary between the Yangtze Platform and South China Fold system. In addition to these. there are axial line of high resistivity F_4 of Longmenshan—Panxi tectonic zone and F_5 of Altay—Tianshan. they are possibly the deep boundary between Qing—Zang Plateau and Yangtze Platform. and the deep boundary of Junggar block.

DEPTH VARIATION OF LOW RESISTIVITY LAYER IN THE CRUST.

In the Chinese mainland the low resistivity layer in crust was found in many regions except for part of regions such as South China Fold system, Yangtze Platform and Songliao block. The top depth of low resistivity layer in crust is generally about 15—30 km with several to more than 10 km of thickness and several to 10 Ω. m of resistivity. Most of upwarping regions of low resistivity layer in crust correspond to the active tectonic areas, where the geothermal temperature is high (Table 1). Besides, the upwarping regions of low resistivity layer in crust correspond to the upwarping regions of Moho (Figure 3) and low resistivity layer in upper mantle mostly (Figure 4). The focuses of shallow earthquake are generally located over low resistivity layers in crust (Figure 5).

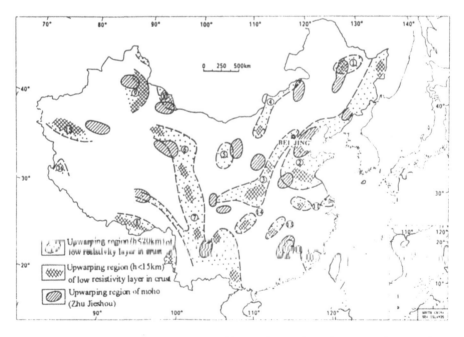

Figure 3. Distribution of upwarping regions of low resistivity layer in crust and upwarping regions of Moho, in which: ①. Northern Songliao block, ②. Eastern fringe of North China Platform, ③. Fen—Wei graben, ④. Eren—Hohhot—Dongshen, ⑤. Northern part of South——North earthquake zone, ⑥. Qilianshan, ⑦. Songpan — Garze — Kangdian area, ⑧. Yarlung Zangbo jiang — Tengchong, ⑨. Tianshan — Junggar, ⑩. Beishan, ⑪. Lower Yangtze, ⑫. Quanzhou—Ganzhou, ⑬. Dongting basin, ⑭. Jianshi—Enshi, ⑮. Northern part of west Kunlun, ⑯. Banggong Co—Gar area.

Table 1. Upwarping regions of low resistivity layer (LRL) in crust

	Region	Depth of LRL in crust (km)	Remarks
1.	Northern Songliao block	11—24	High geothermal and high heat flow
2.	Eastern fringe of North China Platform	14—20	High geothermal, uplifted LRL in upper mantle, depth of earthquake focus: 10—24 km.
3.	Fen—Wei graben	13—20	Uplifted LRL in upper mantle, depth of shallow earthquake focuses: 10—20 km.
4.	Eren—Hohhot—Dongshen	12—22	High geothermal.
5.	Northern part of South—North earthquake zone	14—21	Uplifted LRL in upper mantle and Moho, depth of shallow earthquake focuses: 10—30 km.
6.	Qilianshan	11—15	Low resistivity region
7.	Songpan—Garze—Kangdian	11—20	Earthquake region, focal depth of shallow earthquakes about 20 km.
8.	Yarlung Zangbo jian—Tengchong	11—17	High geothermal, uplifted LRL in upper mantle.
9.	Tianshan—Junggar	13—20	
10.	Beishan	12—24	Upwarping region of Moho.
11.	Lower Yangtze	11—20	
12.	Quanzhou—Ganzhou	14—22	Earthquake region, high geothermal.
13.	Dongting basin	15—19	High geothermal.
14.	Jianshi—Enshi basin	11—16	
15.	North of west Kunlun	10—12	High geothermal.
16.	Banggong Co—Gar	15—18	High geothermal.

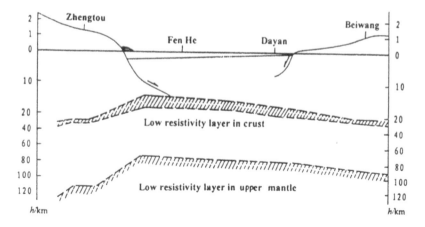

Figure 4. Correspondence between low resistivity layer in crust and low resistivity layer in upper mantle in region of Fen–Wei graben[5].

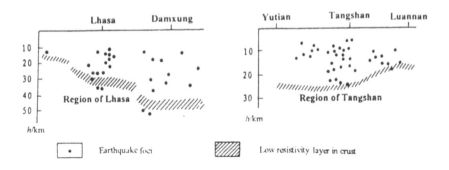

Figure 5. The low resistivity layer in crust and distribution of shallow earthquake focuses in the region of Lhasa and in the region of Tangshan[10].

DEPTH VARIATION OF LOW RESISTIVITY LAYER IN UPPER MANTLE.

In the Chinese mainland the top depth of low resistivity layer in upper mantle varies greatly, from shallowest $50-60$ km to deepest more than 200 km, the average depth is $100-120$ km. In general the depth of low resistivity layer in upper mantle is shallower in eastern China, and deeper in western China (Figure 6). There is a great northeast–southwest trending upwarping zone of low re-

sistivity layer in upper mantle from Songliao through eastern North China Platform towards southwest along Fen — Wei graben, to west fringe of Yangtze
Platform. It is also shown in Figure 1 (geoelectrical structure at 90 km
depth).

Generally, it is said that the low resistivity layer in upper mantle is the underlying layer beneath lithosphere. Based on the depth of low resistivity layer in upper mantle can be ascertained the thickness of lithosphere, and carried out re-
gionalization of lithosphere in Chinese mainland. The region where the thickness of lithosphere is less than 100 km is defined as thin—down region of lithosphere, and the region where the thickness is more than 120 km is defined as
thickening region. Table 2 shows the thickness of lithosphere in China.

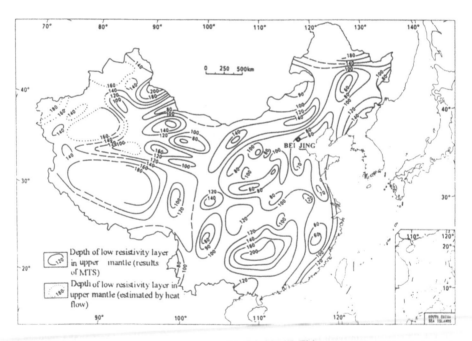

Figure 6. Depth of low resistivity layer in upper mantle of Chinese mainland (depth in km)

The thin—down region of lithosphere mostly corresponds with the high heat
flow region (Figure 7), and most strong earthquakes are distributed in the thin
down regions of lithosphere. It means that the thin — down regions of lithosphere have property of active tectonic zone. In some reference materials it is
proved that large scale endogenic metallogenic provinces are distributed in the
thin-down regions or in the gradient zones of lithosphere thickness[30, 31]. So, the
undulation of low resistivity layer in upper mantle has important meaning for
the mineral prognosis.

Table 2. The thickness of lithosphere in China.

	Thin—down region of lithosphere	Minimun thickness (km)
1.	Songliao—North China region.	60
2.	Eastern fringe of Yangtze Platform and east part of South China Fold system.	50
3.	Western fringe of Ordos block.	67
4.	Fen—Wei graben	55
5.	Kangdian upwarping region	80
6.	Tengchong Folding zone	63
7.	Jinshajiang—Yushu region	92
8.	Eastern Kunlun	87
9.	Qilian Folding zone.	55
10.	Tianshan Folding zone.	64
11.	Eren—Dong Ujimqin zone.	90

	Thickening region of lithosphere	Maximun thickness (km)
1.	Dahinggan—Nei Mongol Folding zone.	170
2.	Southern fringe of Yangtze Platform —Downwarping region of South China Fold system	>200
3.	Youjiang Folding zone	167
4.	Sichuan basin	129
5.	Songpan—Garze region	148
6.	Qing—Zang Plateau	190
7.	Qaidam basin	190
8.	Eastern fringe of Tarim Platform	196
9.	Tarim Platform.	>200
10.	Junggar basin	200
11.	Eastern Liaoning	140

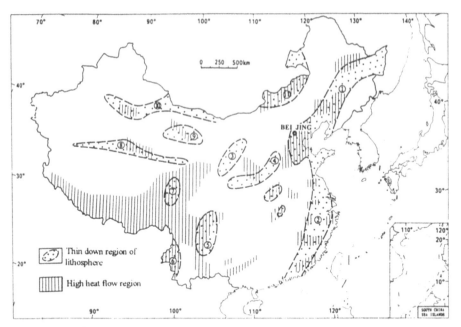

Figure 7. Distribution of thin—down regions of lithosphere and high heat flow regions (Jin Xin)

CONCLUSION

At 90 km depth in the upper mantle a massive low resistivity anomaly is discovered, which run from northeast to southwest from Songliao basin to west fringe of Yangtze Platform. It is inferred that the low resistivity anomaly is an upwarping zone of upper mantle.

The upper mantle from 90 km to 150 km depth in China consists of plastic " soft blocks" and compact "hard blocks" and is not a simple bedded structure. In China at depth of 150 km the mantle material in the majority of areas is in the partial melting state and with low resistivity, only a few regions with high resistivity, they form some "tectonic lines", which are possibly the deep boundaries of geotectonic units.

In most recent active tectonic regions we discovered low resistivity layers in crust, which are buried at shallower depths, less than 20—25 km in general.

A great northeast—southwest trending upwarping zone of low resistivity layer is discovered in the upper mantle, from Songliao to west fringe of Yangtze Platform.

The average thickness of lithosphere in Chinese mainland is 100—120 km, and is smaller in the east than in the west. The thin-down regions of lithosphere are situated mostly at active tectonic regions and correspond with high heat flow and strong earthquake regions.

REFERENCES

1. Yuan Xuecheng, Li Li, Jin Guoyuan. Deep magnetotelluric sounding in the Lhozha—Yang-bajin area, Tibet. *Acta Geologica Sinica*, 59; 25—32 (1987).
2. Li Li, Jin Guoyuan. The electrical charactristic and thermal state of the lithosphere in area Xizang (Tibet), in: *Bulletin of the Institute of Geophysical and Geochemical Exploration*, 2; Beijing: Geological Publishing House, 121—128 (1987).
3. Li Li, Jin Guoyuan. Magnetotelluric sounding study of crust and upper mantle in the Panxi "rift zone" and the Longmenshan faulted zone, *Geophysical and Geochemical Exploration*, 11; 161—169 (1987).
4. Chen Husheng. Comprehensive Geophysical survey of HQ—13 Line in the Lower Yangtze reaches and its geological significance, *Oil and Gas Geology*, 9; 211—222 (1989).
5. Xing Jishan, Yiao Dainqun, Li Mimg. Structural features of graben system in Shanxi discussed on the basis of geophysical data, *Shanxi Geology*, 4; 95—108 (1989).
6. Jiang Hongkan, Zhan Shuangqing, Wang Hongxun. Observations of magnetotelluric sounding performed along Shiyan—Luoyang profile, *Geophysical and Geochemical Exploration*, 14; 285—291 (1990).
7. Jiang Hongkan, Zhan Shuangqing, Wang Hongxun. A preliminary discussion on deep geology along the Macheng—Jiugongshan profile in Hubei province, *Geophysical and Geochemical Exploration*, 14; 357—364 (1990).
8. Zhan Qi. Application of MT in the Northeast of Tarim Basin and its preliminary geological results, *Earth Science*, Supplement 15; 97—106 (1990).
9. Luo Zhiqong. A study of magnetotelluric results in the regional profile from Kaifeng to Lingbi, *Earth Science*, Supplement 15; 87—95 (1990).
10. Zhang Shengye, Ouyang Changding. The application and investigation of magentotellurc sounding for determining regional geoelectric structure in Central Hubei, *Earth Science*, Supplement 15; 79—85 (1990).
11. Li Li, Jin Guoyuan, Liu Changwang. Geoelectrical characteristic of Koktokay—Aksay profile, *Annual of the Chinese Geophysical Society*; Beijing: Seismological press, 49 (1992).
12. Jian Hongkan, Zhan Shuangqing, Wang Hongxun. The deep geoelectrical characteristics in Dazhu Sichuan—Quanzhou Fujian. *Acta Geophysica Sinica*, 35; 214—222 (1992).
13. Gu Qun, Sun Jie, Shi Shulin. The features of high electrical conductivity layer in North China and Northwest China regions, *Seismology and Geology*, 2; 21—29 (1980).
14. National Seismological Bureau. *Geophysical results of crust and upper mantle in China*; Beijing: Seismological press, 271—285 (1986).
15. Wu Guangyao, Zeng Luhai. Application of magnentotelluric sounding in Zhangzhou geothermal area in Fujian, *Earth Science*, 13; 325—333 (1988).
16. Zhang Yunlin, An Haijing, Liu Xiaolin. The relation of lateral change of electric property in crust—upper mantle with great earthquake in partial region of Northwest China, *Seismology and Geology*, 11; 35—45 (1989).
17. Sun Jie, Xu Changfang, Jiang Zhao. The electrical structure of the crust and upper mantle in the west part of Yunnan province and its relation to crustal tectonics, *Seismology and Geology*, 11; 35—45 (1989).
18. Gao Wen, Jiang Bangben, Bai Denghai. Magnetotelluric sounding in the Xingtai earthquake area of Eastern China, *Acta Geophysica Sinica*, 33; 291—297 (1990).
19. Deng Qianhui, Liu Guodong, Liu Jinhan. The observation of magnetotelluric and electrical characteristics of the crust—upper mantle along Hubei Xiangfan—Fujian Luoyuan profile, *Seismology and Geology*, 12; 149—159 (1990).
20. Shi Shulin, Xu Chang tang, Wang Jijun. Study on electric conductivity of the deep earth along Yixian, Liaoning, to Dong—Ujimqin, Nei Mongol profile, *Seismology and Geology*, 13; 115—125 (1991).
21. Kong Xiangru, Liu Shijie, Zhang Jianjun. Magnetotelluric sounding studies in the eastern

area of Fujian province. *Acta Geophysica Sinica.* 34.*724—735 (1991).

22. Bai Denghai, Zhang Li, Kong Xiangru. A magnetotelluric study of the paleozoic collision zone in the east of Inner—Mongolia, *Acta Geophysica Sinica.* 36.*773—783 (1993).

23. Qin Guoqing, Chen Jinhui, Liu Dajian. The characteristics of the electrical structure of the crust and upper mantle in the region of the Kunlun and the Karakorum mountains, *Acta Geophysica Sinica.* 37.*193—198 (1994).

24. Belyavskiy V., A geoelectrical model of Tian—Shan, Physics of Earth, 1, 3—13 (1995).

25. Adam A., Geothermal effects in the formation of electrically conducting zones and temperature distribution in the Earth, *Physics of Earth and Planetary Interiors,* 17, 21—28 (1978).

26. Vanyan L.. The international symposium"Nature of electrical conduction of the Earth's crust", *Physics of Earth,* 1, 95—96 (1995).

27. Ramaletgynov A.. *Model of electrical conduction of lithosphere,* Leningrad: Nauka, 125—147 (1990).

28. An Zhenchang, Ma Shizhuang, Tan Donghai. Spherical cap harmonic analysis of satellite magnetic anomaly in China and its adjacent region, *Annual of the Chinese Geophysical Society.*Beijing: Seismological press, 39 (1991).

29. Ma Xingyuan. *Lithospherlic dynamics atlas of China.* Beijing: China Cartographic Publishing House, 60 (1989).

30. Sheglov A.. Endogenic metallogeny and the tectosphere, *Geotectonic,* 5, 9—16 (1990).

31. Guo Wenkui, Liu Menggeng. *Metallogenic map of endogenic ore deposists of China.*Beijing: Cartographic Publishing House, (1987).

32. Huang Jiqing, Ren Jishun, Jian Chunfa. *Geotectonics of China and its evolution.*Beijing: Science Press 29—65 (1981).

* *in chinese*

Proc. 30th Int'l Geol. Congr., Vol. 20, pp. 95-101
Liu (Ed.)
© VSP 1997

Ultrasonic Velocity of Various Rocks at High-Pressure and High-Temperature Conditions

XU JI-AN
Institute of Earth Sciences, Academia Sinica, Nankang PO Box 1-55, Taipei City, Taiwan 11529, China

XIE HONGSEN AND ZHANG YUEMING
Institute of Geochemistry, Chinese Academy of Sciences, Guiyang 550002, China

Abstract

A review on the recent progress of our ultrasonic velocity measurements at simultaneous high-temperature and high-pressure conditions is given. Using the pulse transmission technique, the compressional ultrasonic velocity (Vp) of various rocks at simultaneous high-pressure and high-temperature conditions up to 5.5GPa and 1500℃ has been measured. According to our experimental results, in a microcrack-free sample, the observed compressional velocity (Vp) will be sharply reduced by a large amount upon initial compression. This effect gradually decreases with pressure and totally disappears at pressures above 2.5 GPa. Accordingly at ambient conditions, the Vp in a basalt sample without microcracks should be 6.856 km/s, much higher than the value, measured from the sample with microcracks, of 6.044 km/s which was the normal accepted value in handbooks. Hence we believe that microcracks exist in most of basalt samples which have been tested ever.

In the simultaneous high-temperature and high-pressure experiments, a softening phenomenon occurs at a special temperature θ_x which indicates a kind of softening for rocks. It is noteworthy that the θ_x corresponds to the glass transition temperature θ_g for glass samples. Therefore the rocks behave similarly as glass at high temperature. In addition, basalt was found to transfer to eclogite above 3.5 GPa and 500℃ which might be a major process between the subduction zone and mantle in the deep interior of the Earth.

Keywords: Ultrasonic velocity, High pressure, High temperature, Basalt, Eclogite

INTRODUCTION

The information of the ultrasonic velocity on various Earth's materials at simultaneous high-temperature and high-pressure conditions is meaningful for understanding the structure and status in the core, mantle, and the properties in the low velocity zone and other discontinuity layers in the interior of the Earth. Such information provides an experimental basis for the mechanism studies of various geological disasters. Recently a system for such studies up to 6.5 GPa and 1500℃ simultaneously has been set up [1]. In our previous measurement, we found:

(1) The observed velocities depend upon the existing microcracks in the sample, hence the velocities are not intrinsic parameters. However, these microcracks close and their influence on velocities disappears at pressures above 2.5 GPa for all tested rocks, thus the observed velocity above 2.5 GPa represents the actual velocity under pressures;

(2) A softening temperature (denoted as θ_x in this paper) exists in all of the tested glass and crystalline materials, although it is corresponding to strain point Ts for glasses . The physical reason of such a softening for the crystalline material is still unknown;

(3) The softening temperature θ_x is independent of the microcracks in the tested sample, and it shifts with pressure.

In the present work, the recent results on the velocity measurement on various rocks up to 5.5 GPa and 1500°C under simultaneous high-temperature and high-pressure conditions are discussed.

EXPERIMENTS

The experiments were performed in a high-temperature and high-pressure cell in a YJ-3000 press (maximum force: 3000 tons) at the Institute of Geochemistry, Chinese Academy of Sciences, Guiyang, China. The experimental details were described else where [1]. The samples tested were pyrophyllite, kimberlite and basalt. The rock was a condensed body with a uniform distribution of various minerals and without obvious pores, hence, the observed compressional velocities at ambient

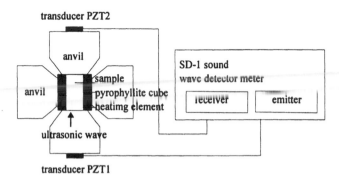

Figure 1 Experimental setup.

conditions, $(Vp)o$, were either only slightly different (in kimberlite) or identical (for basalt). In the latter case, it was (6.055 ± 0.010 km/s) in the samples which were cut in different orientations.

The experimental setup was shown in Figure 1. The samples were cut to 33 mm in length and 12 mm in diameter, and fitted into a sample hole in the pyrophyllite cubes along with the heating elements (three stainless steel foils). The ultrasonic vibration was produced from a transducer (PZT1) located in the back of the lower anvil, and received by the other transducer (PZT2) located in the back of the upper anvil, the velocity was determined from the travel time in the sample. When the anvils squeezed together and heating electrical current passed through the stainless steel foils outside of the sample, the simultaneous high-pressure and high-temperature conditions were generated in the pyrophyllite cube, although the temperature in the sample was not uniform but with a special distribution, and the ultrasonic velocity measurement at such conditions could be performed. The highest temperature across the sample in its z-direction, achieved in the middle of sample and denoted as θ_{max}, which was calibrated by the heating power W. The temperature error was of the order of 5 - 20℃. The accuracy of pressure values was believed to be within 0.2 - 0.5 GPa and the error of velocity was less than 6% which was due to the changes of the velocity and the length in both upper and lower anvils at different temperatures and pressures.

COMPRESSIONAL VELOCITY Vp AT HIGH PRESSURES

In our ultrasonic measurements, an abnormal behavior is observed in the glass samples in the initial stage of compression (less than 0.5 GPa). The velocity Vp suddenly decreased by a large amount. This abnormal behavior is due to the

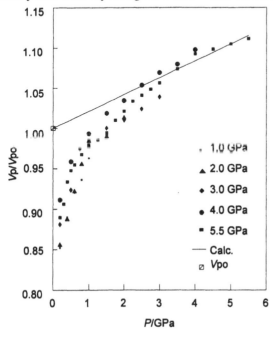

Legend:
- 1.0 GPa
- ▲ 2.0 GPa
- ♦ 3.0 GPa
- ● 4.0 GPa
- ■ 5.5 GPa
- —— Calc.
- ⊠ Vpo

Figure 2 Compressional velocity Vp/Vpo for kimberlite as a function of pressure (the straight line is the fitting curve of the experimental data above 2.5 GPa in the present study).

microcracks created during the initial compression under non-hydrostatic condition, indeed, we also noticed sound emission from the sample[1]. Same result was found in kimberlite as shown in Figure 2. Obviously, the initial velocity $(Vp)o$ can be obtained when extrapolating the fitted line from the data above 2.5 GPa to ambient conditions.

A different case of basalt is shown in Figure 3. As described above, the $(Vp)o$ in a basalt sample without microcracks, was expected to be 6.856 km/s (extrapolated value from the data above 2.5 GPa), but the observed $(Vp)o$ value was about 6.055 km/s. Hence, we believe that there were huge numbers of microcracks in the original sample, and as expected, the they were closed at around 2.5 GPa. Obviously, the 10% velocity variation could be produced in the samples with different degree of microcracks.

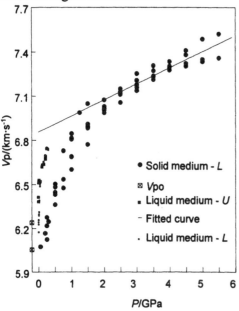

Figure 3   The compressional velocity of basalt Vp at various pressures at room temperature. The solid square points obtained at a hydrostatic environment condition (L - Loading, U - Unloading), the observed pressure dependence of Vp in such a condition is a little larger than the one obtained from the non-hydrostatic environment.

The square experimental points shown in Figure 3 were measured at a hydrostatic condition in loading and unloading process up to 0.2 GPa and room temperature by a pulse transmission method[?]. In such an environment, there should be no additional micro-cracks produced in the loading process, the pressure dependence of compressional velocity (dVp/dP) was even larger than the value measured in the non-hydrostatic environments. The results lead same conclusion: there were a huge number of micro-cracks in the original basalt samples, the microcracks decreased in the loading process, and the velocity change in such a sample at initial compression was caused by a dual effects: 1. the actual compression; and 2. enclosing of microcracks. Hence, the pressure dependence (dVp/dP) could be extraordinarity large in its initial compression due to the latter effect which was eliminated above 2.5 GPa.

Most compressional velocity Vp of basalt reported ever in handbooks was less than 6.0 km/s[3], therefore, we believe that most basalt samples tested involved microcracks.   They can not represent the actual velocity in deep interior of the Earth.

THE Vp - θ_{max} RELATIONSHIPS AT HIGH PRESSURES:

Softening

In the relationships of velocity vs. temperature of rocks, a normal behavior (Figure 4) was shown wich a softening phenomenon at softening temperature θ_x for all of the experiments on basalt below 3.0 GPa as well as pyrophyllite and kimberlite [1]. The softening temperature θ_x shifts with pressure.   After the experimental run, the Vp returns to its original (or a little lower) value.   Such a softening phenomenon might appear in all Earth materials, therefore, it provides a possible interpretation of the low-velocity zone in the mantle.

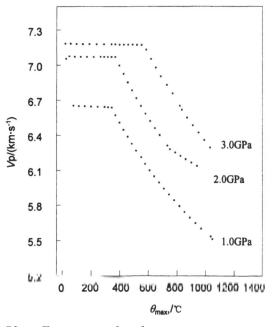

Figure 4   The relationships between the compressional velocity of basalt Vp and temperature at various pressures below 3.0 GPa.   The similar experimental curves can be obtained for pyrophyllite and kimberlite [1].

Phase Transition in basalt

Another type of relationships between Vp vs. θ_{max} is shown in Figure 5.   A velocity increase was shown clearly above 400° - 600°C at different pressures above 3.5 GPa and after the first experimental run the Vp did not return to its original, but a much higher value.   The microcracks had been closed when the applied pressure was above 2.5 GPa, hence such an increase should not depend on the disappearance of microcracks.   After an experimental run, the serpentine in basalt disappeared and garnets (mainly grossular) produced.   Therefore we believe that the transition from basalt to eclogite should be responsible for such a behavior.

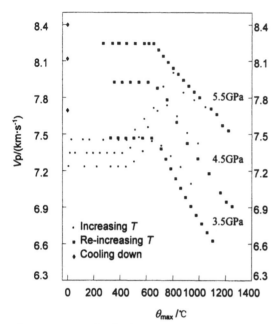

Figure 5 The relationships between the compressional velocity of basalt Vp and temperature at various pressures above 3.5 GPa. The increase of velocities corresponds to a phase transition from basalt to eclogite. After the transition, the samples at various pressures were cooled down to around 250℃, then heated up to high temperatures, the velocities in the processes are shown as the empty-square data points in this figure. The normal Vp - pattern similar with figure 4 are shown, they correspond to the curves of eclogite.

The structure of eclogite is much denser than basalt, and a higher velocity is expected. After the first-time experimental run, the eclogite formed, the $(Vp)o$ returned a higher value than its original, and in the subsequent heating runs, the relationships of Vp vs. θ_{max} are similar to the normal cases, the observed θx are the softening temperatures of eclogite at various pressures..

Because the transition from basalt to eclogite is the major process between the subduction zone and mantle, this transition provides a possible source of deep earthquakes. The detail studies on such a transition would be very useful in the Earth science. It is very interesting to know what effect the water in the sample had ? Was dehydration involved in the transition ? They would be the subjects in our future study.

The authors gratefully acknowledge the editorial help from Mr. Xu, Zuming in the Institute of Geochemistry, Chinese Academy of Sciences and the help during the experiments from the Institute of Geochemistry, Chinese Academy of Sciences, Guiyang and the Institute of Earth Sciences, Academia Sinica, Taipei and thank for the support from the National Natural Scientific Foundation, Beijing and the National Science Council, Taipei, through NSC-84-0202-001-015.

REFERENCES

1. Xu.J, Zhang.Y, Hou.W, Xu.H, et al., Ultrasonic wave speed measurements at high-temperature and high-pressure for window-glass, pyrophyllite and kimberlite up to 1400℃ and 5.5 GPa, *High-Temp. and High-Pres.* **26**, 375-384 (1994).

2. Xie.H, Zhang.Y, Hou.W, et al., Proceedings of the International Conference and 3rd Sino-French Symposium on Active Collision in Taiwan, Edited by S. Chien, Taipei, Taiwan University Press, 301-307(1995).

3. N. I. Christensen, Seismic Velocities, *Practical handbook of physical properties of rocks and minerals,* Ed. R. S. Carmichael, Inc., Boca Raton, CRC Press, 429 - 546 (1989).

Proc. 30th Int'l Geol. Congr., Vol. 20, pp. 102-111
Liu (Ed.)
© VSP 1997

The Essence of Techniques and Methodology of Geophysics and Remote Sensingfor Exploration of Sylvinite in China

ZHANG YU-JUN, LI CHANG-GUO

Center of AeroGeophysics & Remote Sensing (AGRS) MGMR, 29 XueYuan Road, Beijing, 100083, P.R.C.

Abstract
Owing to the severe shortage of the potash fertilizer in China great attention has been paid to the study of new methods, including gamma-spectrometric well-logging, airborne gamma-spectrometry, remote sensing andgravimetry, for exploration of potash salt during the past more than 30 years. This paper shortly summarizes the achievements of our efforts dedicated to the research work in this field.

The gammaspectrometric well-logging method and bore-hole gammaspectrometer were proposed and developed by us in 1963, and several times updated up to the eighties. The gammaspectrometric well-logging has been applied in 6 provinces successfully. A bore-hole radioactivity-spectra interactive interpretation software system (HRSIS) was developed in 1995, which consists of 5 groups of functional programs: the preprocess, the quantitative interpretation, the trend analysis, the graphic display and the integrated interpretation.

The high sensitivity airborne survey has been conducted since 1984 for investigating chiefly the evaporite-type potash deposits in the western China. 4 parameters are usually recorded: K, Th, U and magnetic field. The airborne multichannel gamma-spectra processing software package (AMGPSP) and the aerogeophysical image processing software package (AGIPSP) were developed in the eighties for preprocessing, spectra-stabilization and image processing of aero-gamma-spectra data.

Since the middle eighties the digital image processing techniques have been used for integrated data interpretation, including geophysical and remote-sensed MSS & TM data, in studying the evaporite-type sylvinite-bearing basins in the west China.

At the beginning of 1993 a very special software package for 3D imagery of gravitational field (3DIG) was developed by us. It constructs the 3D matrix of a gravitational field and then displays it in every variety of form to discover the internal features of the given gravitational field.

Keywords: Potash salt, Sylvinite, Bore-hole gammaspectrometer, Airborne gammaspectrometry, Remote sensing, 3D imagery of gravitational field.

GAMMASPECTRUMETRIC WELL-LOGGING METHOD

According to the statistics more than 50% of the total world reserve of sylvinite were discovered in bore-holes. For revelation of sylvine and differentiation of it from the argillaceous thorium-uranium-bearing rocks the gammaspectrometric well-logging method was studied and the bore-hole gammaspectrometer was developed since 1963 and 5 times updated up to the eighties.

It is the natural radioactive isotope K40 that lays the physical base for this method. Isotope K40 has low isotope abundance 0.0119% and radiation coefficient of gamma photon 11.6% with middling energy 1.46, suitable for detection and spectra analysis.

The 4 channel bore-hole gammaspectrometer, developed by us, registers thegamma-ray spectra by a cylindrical NaI(sodium iodide) crystal. The 4 channels are: the channel PK (differential channel for K40 peak of 1.46MeV. The energythreshold is 1.40--1.52MeV); the channel LI (left integral channel withenergy threshold => 1.30MeV); the channel RI (right integral channel withenergy threshold => 1.60MeV) and the channel TC (total count channel, its energy threshold is => 0.2MeV). The following technical points were studied for construction of the bore-hole gammaspectrometer:

* Sensitivity. Because of the low isotope abundance and gamma radiation coefficient it was necessary to raise the sensitivity of the bore-hole gammaspectrometer by increasing the length of the NaI detector up to 100--150mm. Its diameter is limited by the inside diameter of the hole and can not be > 50mm.

* Impulse transfer. In order to keep the resolution the impulses (from photomultiplier after preamplifier) must be transferred from the down hole unit through a cable, longer than 3600m, without serious deformation in shape and in amplitude. A voltage-current type transfer was accepted. The resolution for Cs137 after 3600m cable transfer kept the same level of resolution (10%) as without the long cable.

* Stability. The vertical temperature gradient of the earth is about 1 degree Centigrade for each 33m, so at the depth 3500m the environmental temperature in bore-hole can be as high as 100 degrees Centigrade. For the reliable stabilization of the spectrometer the following 3 measures were adopted: the selective use of high temperature NaI crystal and photomultiplier; the compensation of the nega-

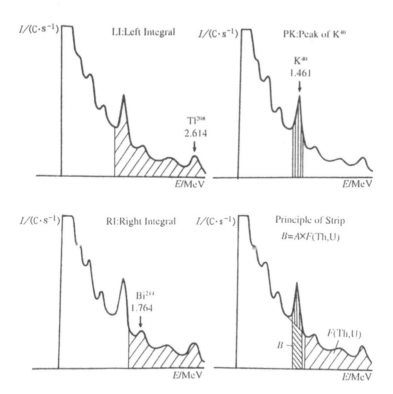

Figure 1. The 4 channel bore-hole gammaspectrometer registers the gamma-ray spectra in the hole by a cylindrical NaI(sodium iodide) crystal. The 4 channels are: the channel PK (differential channel for K40 peak of 1.46MeV. The energy threshold is 1.40--1.52MeV); the channel LI (left integral channel with energy threshold => 1.30MeV); the channel RI (right integral channel with energy threshold => 1.60MeV) and the channel TC (total count channel, its energy threshold is => 0.2MeV).

tive temperature coefficient of the photomultiplier by the positive temperature coefficient of the high voltage source for the photomultiplier; and the use of an autostabilization circuit for isotope Cs137 to keep the spectral peak position.

Any photo-peak of an isotope is added on the compton continuum of photo-peaks with higher energy. It is the theoretical base of the stripping method and the principal component analysis for K, that the background is proportional to the RI (Fig. 1). We have studied to use the principal component analysis and the stripping method for depression of the U-Th interference and thereby to get purer potassium anomaly. The mathematic process of the principal component analysis is shown by expressions (1). The eigen vectors were calculated by statistics.

$$
\left.
\begin{array}{ll}
 & E_i = \sum_{k=1}^{n} L_{ik}X_{ik} \\
\text{Covariance Matrix } C & C = 1/n \cdot XX' \\
\text{Eigen Values } G & |GI\text{-}C| = 0 \\
\text{Eigen Vactors } L & (GI\text{-}C)L = 0
\end{array}
\right\} \tag{1}
$$

where X -- the old variables; E -- the new variables, which are the linear functions of the old variables; I --the unit matrix.

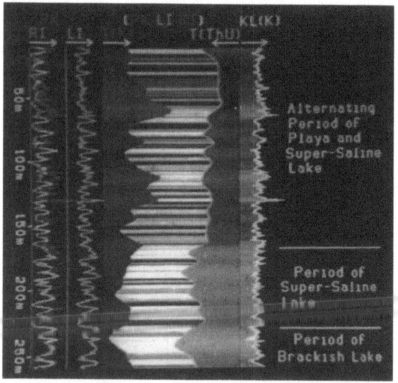

Figure 2. The example of well-logging images for the borehole, located in the very center of the sedimentation activity of KunTeYi Basin in QinHai province.

* On the left there are the original gammaspectrometric logs, scaled by eigen vectors.

* In the middle there is the composite ternary image of PK, LI & RI. The rock salt layers are dark because of the low concentrations of radioactive elements, on the contrary the sylvine or the ischelite containing layers are bright. The argillaceous rocks are bright too due to the higher concentrations of U,Th and K.

* The trend analysis curves for K and Th-U illustrate the step decreasing for Th-U and increasing for K, reflecting the 3 periods of the basin evolution: the period of brackish lake; the period of super-saline lake; the alternating period of playa and super-saline lake.

* The K curve, obtained by the principal component analysis, shows the enriched ischelite layers at depths 60m and 130m.

A bore-hole radioactivity-spectra interpretation software system (HRSIS) was developed in 1995, which consists of 5 groups of functional programs: the preprocess, the quantitative interpretation, the trend analysis, the graphic display and the integrated interpretation. The sixth group is for the HELPs. The HRSIS system was written on PCs with C language. Fig. 2 shows one example of the well-logging images for the bore-hole, located in the very center of the sedimentation activity of KunTeYi Basin in QingHai province. On Fig. 2 the KCl layers are revealed by high LI, high PK, high TC and low RI, besidesthe KCl layers the 3 periods of lake sedimentation are seen clearly too.

Gammaspectrometric well-logging method has been applied in 6 provinces (YunNan, HuBei, SiChuan, ShanDong, XinJing and QingHai) successfully. The sylvinite-bearing layers with thickness => 0.5m and KCl concentration => 2% can be detected by bore-hole gammaspectrometer reliably in MengYeJing and other salt deposits. The policy of "Exploration for both Oil and Sylvite Concurrently" was made by Ministry of Petroleum of PRC. Following this policy a aphthitalite layer with thickness 0.8m and KCl concentration as high as 16.8% was found in JiangHan oil field using the gammaspectrometric well-logging. In province SiChuan the ischelite layers with thickness of => 0.2m were detected. In 1991 in YunNan province in a new drilling area, when all of the drilling core of the first borehole was lost, only thanks to the gammaspectrometric well logging a big amount of money was saved by avoidance of redrilling.

AIRBORNE-GAMMASPECTROMETRY

The high sensitivity airborne survey has been carried out since 1984 for investigating chiefly the evaporite-type potash deposits in western China. 4 parameters are usually recorded: K, Th, U and magnetic field. It uses both the 4 channel and multichannel (512 channels) airborne gammaspectrometers. The total volume of the prismatic sodium iodide crystals achieve 50 litres. The terrian clearance is about 90m and the survey scale is 1:100000--1:200000 with line- spacing of 1--2km.

The airborne multichannel gamma-spectra processing software package (AMGPSP) and the aerogeophysical image processing software package (AGIPSP) were developed in the eighties in FORTRAN for preprocessing, spectra-stabilization and image processing of aero-gamma-spectra data [1,2]. The areas in western China for investigating potash salt are characterized by very sparse vegetation and mild surface relief and are, therefore, ideal for airborne-radiometry. The dynamic ranges are usually about 0 8% for K, 0--40ppm for Th and 0--15ppm for U.

The digital image processing techniques are of big help for extraction of the geological information in the following aspects: the lithological mapping; the direct detection of potassium; the study of the sedimentation of potash deposits; the extraction of structural traces; and, in some cases, the oil-gas anomalies identification [3,4]. Aero-gamma-spectrometry is one of the few geophysical methods, that can be used for direct detection of a certain element.

In substance, the ternary image of K, Th and U data, as the results of gamma- spectrometry, is a regional geochemical map of the three elements. In integrating with LANDSAT images (TM or MSS) the ternary image of K, Th and U can serve the purpose of lithological mapping by a supervised or unsupervised classification method.

The abundance of potassium oxide in the argillaceous rocks (sandstone, mudstone...) makes the main interference in direct detection of KCl. By analysis of the binary image of K & Th and their scattergram it was discovered that there is a linear correlation between the concentrations of K and Th in argillaceous rocks. A series of image processing techniques, such as the principal component analysis, the transformation between color coordinate systems, the vegetation extraction..., were tested for direct detection of KCl.

Based upon the above mentioned phenomenon we have succeeded in distinguishing the potassium chlorate from potassium oxide, which abounds in the Tertiary sandstone in the NE corner of the survey region in ChaiDaMu basin. More than 10 new anomalies for potassium chlorate were outlined for this survey region, one of them was followed up by groud survey, and the discovered reserve of the solid KCl was increased by 3,245,000t. The KCl bittern is of particular importance for exploration. We've defined the KCl bittern as the surface water with K concentration >= 0.6%.

By a supervised classification we've obtained 2 more classes in addition to the 14 rock classes on the geological map. We've succeeded in distinguishing KCl not only from potassium oxide, but also from KCl with argillaceous.

REMOTE-SENSING

The 7 banded TM data, benefiting considerably from their higher spectral and spatial resolutions, are more helpful for study of potash-bearing evaporite- type basin.

Although the geologic information extraction is hoped to use as more bands as possible, a well-displayed ternary image is always necessary. The best ternary image should contain the most quantity of

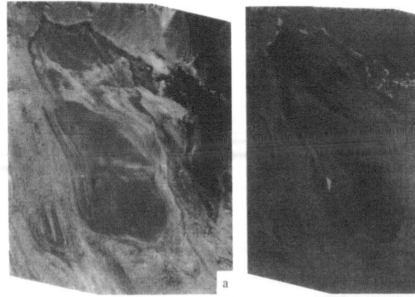

Figure 3. a: The composite image from the 1,4 & 7 bands of TM data clearly shows the general geomorphological landscape of KunTeYi basin in the western China. It is a rhombus shaped close playa with 80 km in length and 30 km in breadth. b: By KL transform of the 1,4,5,6,7 band data the valuable anomaly image was got, where the sylvinite is manifested as bright spots.

information and with the least correlation between the three bands. A composition-correlation-factor Q was established as the measure and the basis for the optimized composition.

$$Q = \sum_{i=1}^{3} S_i / \sum_{i=1}^{3} |R_i| \tag{2}$$

where S_i –is the variance for i band; R_i --is the correlation coefficient.

The best ternary image should have the most Q. For instance, the best ternaryimage was found for KunTeYi basin as the composite of the 1,4 & 7 bands of TMdata. KunTeYi basin is a recent salt lake with a certain importance for explorationof sylvite. It is located in the north-west ChaiDaMu basin in QingHai province.Fig. 3a is the B/W image, which clearly shows the general geomorphological landscape of the basin. It is a rhombus shaped close playa with·80km in lengthand 30km in breadth. The high way goes across the playa, which is surroundedby Arjin mountains in the north and 2 series of anticlines in the east, west and south, some of them are with productive oil fields.

For establishing the anomaly indicator of solid potassium chloride the imagesampling was done. 14 geologic objects were selected on the image and their intensity varying on the 7 banded TM data was analyzed. Fig. 4 illustrates the image sampling results. The solid sylvinite (No.12 & 13) has obvious distinguishable spectral characteristics, providing the basis for interpretation.

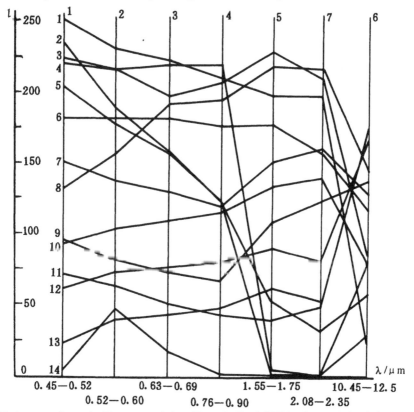

Figure 4. The image sampling results. The curves are the intensities on the 7 banded TM data for 14 typical geologic objects. The solid sylvite (No.12 & 13) has obvious distinguishable spectral characteristics, providing the basis for establishing an indicator for solid potassium chloride.

Based upon the spectra analysis both the KL transform (the principal component analysis) and the level-slice were applied for extraction of the potassium chloride anomalies (Fig. 3b). The 2 methods gave very close results. Besides the detection of KCl, the study of structural features, the analysis of saline deposit properties, and the revelation of material sources were also involved.

GRAVITATIONAL FIELD 3D IMAGERY

The density contrast between the salt layers and the surroundings usually is greater than 0.3g/cm3; and the density contrast between sodium and potassium chlorides is about 0.15g/cm3. So gravity is used traditionally for exploration of salt deposits.

In YunNan province geophysicists used to calculate the differential of the averages of the residual gravity on circumferences with variable radius to evaluate the shape and size of the salt ore body by formula (3).

$$\triangle g^{1\text{-}2} = (\sum_{i=1}^{n} g_{r_1})/n_1 - (\sum_{i=1}^{n} g_{r_2})/n_2 \tag{3}$$

whter g_{r_1}, g_{r_2}--data of the residual gravitational rield on the circumference with radius r_1, r_2. n_1, n_2--numbers of points on circumference with radius r_1, r_2, taken for averaging.

Principally speaking, this differential averaged residual gravith is similar to the first vertical derivative of gravity. Sveral examples of interpretation by this method were veryfied by drilling.

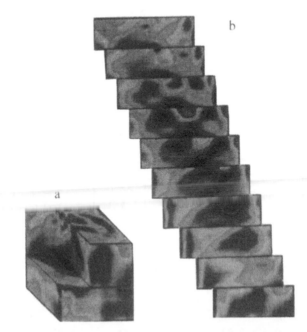

Figure 5. The example of interpretation results by the software system 3DIG. a: Pseudo "CT" image for the 20th local negative anomaly of MengLa gravitational field in province YunNan. b: 10 transections along the profile of the 20th local negative anomaly, showing the internal relationship of the profile with its surrounding environment and giving more detail description of the deformation and the stereo relationship with the faults of the basin.

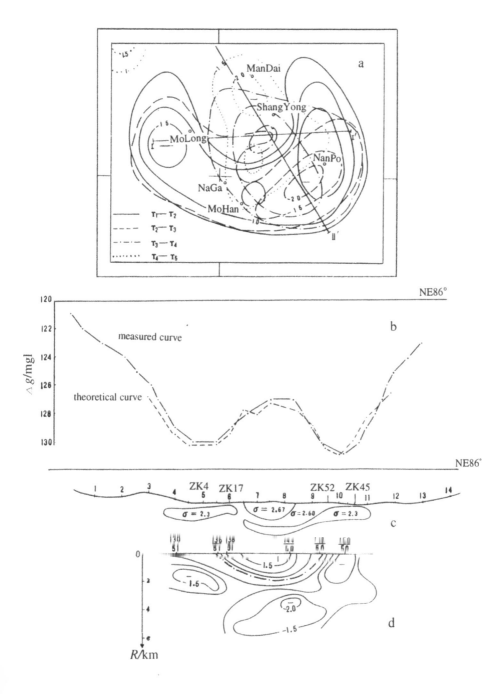

Figure 6. The interpretation results for the 20th local negative anomaly, obtained by geophysicists of YunNan province manually. a: Plan of differentials of the averages of the residual gravity on circumferences with radii r1, r2, r3, r4, r5. b: Profile interpretation. c: Interpreted rock salt body. d:Profile section.

The contribution of our work consists in the 3D imagery of gravitational data ror a better visibility and a hifger rapidity in the interpretation, At the begining of 1993 a very special software package for 3D imagery ofgravitational field(3DIG)was developed by us in FORTRAN. It uses the upward continuation, the differntial averaged residual gravity on circumferences with variable radii, or other methods to construct the 3D matrix of a gravitational field and then displays it in every variety of form.

This software package was applied for reevaluation of the salt-bearing MengLa gravitational field in SW China. It is a gravitational field with dynamic range from -3.5mgl to 2.9mgl. There are 20 local negative anomalies. Each one of the local anomalies was studied by 3D imagery in form both of the stack-profiles (or stack-sections) and of pseudo "CT", showing the internal relationship of the profile with its surrounding environment and its 10 transections along the profile, giving more detail descrip tion of the deformation and the stereo relationship with the faults of the basin. Fig. 5 shows this kind of images for the 20th local negative anomaly, for which the comparison was made by manual calculations. Fig. 5 is the result of manual calculation. Both Fig. 5 and 6 discover the internal features of the 20th local negative anomaly. Fig. 7 demonstrates the main section across the whole area from the N to the S and then to the SE in form of profile with isolines. It was cut into 4 parts because of the line display limit.

The depth scale could be obtained by correlation analysis (for example by the least square fitting), using the drilling results and the section data. Once the approximate depth scale has been got, the software can easily calculate the depth, the thickness and the reserve of potash or sodium salt.

Figure 7. The main section across the whole MengLa gravitational field from the N to the S and then to the SE in form of profile with isolines. It was cut into 4 parts because of the line display limit.

Acknowledgements

We thank Mi Shuang-Jie, Li Show-Tian, Zhang Ming-Lun, Chen Yun-Long, Li Shu-Yi,Chen Xian-Yao and others, who have taken part in the development of the bore-hole gammaspectrometer and the study of the method. Thanks also due to the field geophysicists of province YunNan, who have worked for gravimetry on salt. We are grateful to Zhu Ye-E, Shi Dian-Lin and Xie Xing for their work on software developing and to Xu Dong-Chen for the airborne geophysical survey in QingHai province. We are also grateful to Gao Yong-Jin for her support with S600 image system, to Yang Xing-Hong and Li-Yi for their help in printing of the graphs and the text.

REFERENCES

1. Zhang Yu-Jun, Wang Nai-Dong and Zhang Zhi-Min. A software package for processing of data of airborne gamma-ray multichannel spectrometer,*Computing Techniques for Geophysical and Geochemical Exploration.* **11**, 11-21 (1989).
2. Zhang Yu-Jun.The study of image restoration techniques for aerial radiometric data,*Chinese Journal of Geophysics,* **33**, 405-412(1990).
3. Zhang Yu-Jun. A study of aero-magnetic and radiometric anomalies in some oil-gas field by image processing techniques, *Chinese Journal of Geophysics,* **37**, 505-515(1994).
4. Zhang Yu-Jun. Digital image processing of airborne radiometric and magnetic data from central Chai DaMu Basin, In: *An Overview of Exploration Geophysics in China.* Tulsa, American Society of Exploration Geophysicists. 517- 535(1989).

Proc. 30th Int'l Geol. Congr., Vol. 20, pp. 112-120
Liu (Ed.)
© VSP 1997

Geophysical Analysis of Main Minerogenetic Belts in China

YANG WENCAI

Institute of Geology, Beijing, 100037, China

HOU ZUNZE

Institute of Geophysical and Geochemical Exploration, Langfang, 065000, China

Abstract

Minerogenetic domains might be caused by inhomogeneous dynamic and chemical activities during the creation of the original crust. The measurements of abundance of chemical elements have also provided evidence for the inhomogeneous mass distribution. Based on study of deep and superdeep continental drilling, the GGTs and regional geophysical maps, we find that these domains are characterized by higher crustal average density which can be estimated from gravity data after using new data processing and inversion techniques, such as the wavelet transform and multi-scale analysis. By performing the multi-scale analysis of Bouguer gravity field of China, and comparing with geological findings, a prediction of main mineralization zones in the area are presented.

Keywords: Wavelet transform, Multi-scale analysis, Gravity field of China, Giant ore deposits.

INTRODUCTION

This paper is aimed at prediction of giant heavy metal ore deposits based on geophysical data. For many years geophysicists have been applying some indirect criterion to locate ore deposits. For instance, they use geophysical anomalies to map faults and suggest that fault zones might be possible base-metal minerogenetic zones. These kinds of interpretation often produce some ideas for exploration of ore bodies, but not so useful for prediction of giant ore deposits, giant ore deposits are very rare and the probability of forming them are very small, while the probability of finding faults is large. We have to find some new geophysical characteristics which are directly related to minerogenetism of the giant ore deposits. A suggestion comes from geochemical investigations showing that the abundance of base-metal elements are very rich around giant ore deposits in many areas. We think that the masses, as well as their density, in the uppermost crust are distributed inhomogeneously. It may give some geophysical indications for locating minerogenetic domains of the base-metal deposits.

From continental scientific drilling data obtained at Kola, Uralsk, Muruntau and other places, we find that most of these minerogenetic belts are characterized by high density distribution in the uppermost crust, say greater than 2800 kg/m³; and the P-wave velocity

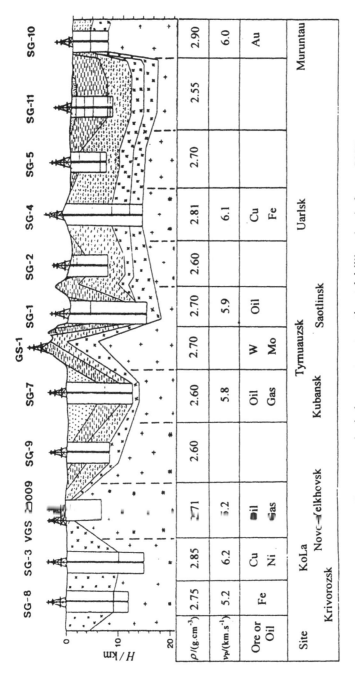

Site												
Ore or Oil	Fe	Cu Ni	Oil Gas		Oil Gas	W Mo	Oil		Cu Fe			Au
$v_P/(km.s^{-1})$	5.2	6.2	5.2		5.8	5.9			6.1			6.0
$\rho/(g.cm^{-3})$	2.75	2.85	2.71	2.60	2.60	2.70	2.70	2.60	2.81	2.70	2.55	2.90

Krivorozsk   KoLa   Novo-Telkhovsk   Kubansk   Tymuauzsk   Saotlinsk   Uarlsk   Muruntau

Figure 1. Upper crust physical parameters at continental drilling sites in former SSSR, showing higher density and P-wave velosity related to giant metal deposits at Kola, Ural and Muruntau.

is also higher in these belts (see Figure 1 for instance). Based on GGT data worldwide we can obtain the same conclusion. For example, the Mandura-Tsibo GGT section shows a very high density zone existing under the giant ore deposit of Paiyunopo. The density there is higher than 2810 kg/m³ and can reach 2950 kg/m³. By the way, we also noticed that the shear wave velocity under these minerogenetic belts in the upper crust can be significantly high, i.e. exceed 3700 m/s, while it is usually about 3500 m/s.

Why the large or giant ore deposits involve with high density uppermost crust? The problem is complicated and hard to answer, but we can consider the fact: the giant ore deposits need great supply of heavy metal masses which usually have higher density than common rock masses. The high density uppermost crust may be initially formed by meteorite impact or volcanic eruptions during the primitive time of the Earth, or even by intrusion and upthrust of the mantel material.

HOW TO LOCATE HIGH-DENSITY UPPERMOST CRUST

The gravity field is generated by the Earth's masses existing at all depth, we usually separated it into the so-call regional anomalies and the residuals which are by no means produced by the uppermost crustal masses. New techniques are required to decompose the Bouguer gravity field into several components which should correspond to the uppermost crust and fluctuation of the Moho Respectively. Thanks to a new mathematical method called the wavelet transform (Daubechies, 1988, 1992; Mallat, 1989), we are able to unfold the gravity field with different scales and sufficient accuracy.

In the wavelet transform theory, the scale decomposition is obtained by dilating and contracting the chosen analyzing wavelet before convoluting it to the gravity field. Mayer and Mallat introduced the concept of multi-scale analysis which gives us a general method to build orthogonal wavelet basis and leads to the implementation of fast wavelet algorithm. This powerful tool has been extended two-dimensional and applied to decomposition of gravity field of China by us (Hou and Yang, 1997). Because the uppermost crust masses usually generate small scale gravity anomalies, they usually correspond to the first order wavelet detail, i.e. the component of smallest scale. As the sources go deeper and deeper, their gravity anomalies will become larger and larger which should be shown in the higher order wavelet details. Figure 2(a) shows a model which consists a three-dimensional high density body located in the upper crust and a fault crossing the Moho, their gravity anomalies are shown in figure 2(b). After applying the multiscale analysis to the gravity field, we obtain the first, the second, the third and the fourth order wavelet details, and named D1, D2, D3 and D4 in Figure 3. Though the first order wavelet detail mainly reflect the uppermost crustal mass distribution, it still contains some influence from relief of the Moho which is the main source of the third order wavelet detail. By subtracting the third order detail from the first order detail (called D1-D3 hereinafter), we find that the difference is just the gravity field caused by uppermost crustal masses, as shown in Figure 2(c).

THE UPPERMOST CRUSTAL DENSITY DISTRIBUTION IN CHINA

More than 10,000 Bouguer gravity data uniformly located in the $1°\times1°$ grids in China are used for multi-scale analysis in this study. The algorithms of two-dimensional wavelet transform and the multi-scale analysis are introduced by the authors (Hou and Yang, 1997). The result of D1-D3 is the maps shown in Figure 4 for positive anomalies and Figure 5 for negative anomalies. In Figure 4 we see the distribution of heavy masses in the uppermost crust, while in Figure 5 we see the lighter masses in the uppermost crust that is more or less coincides to main oil/gas bearing sedimentary basins in China, usually having extension history during Mesozoic era or later.

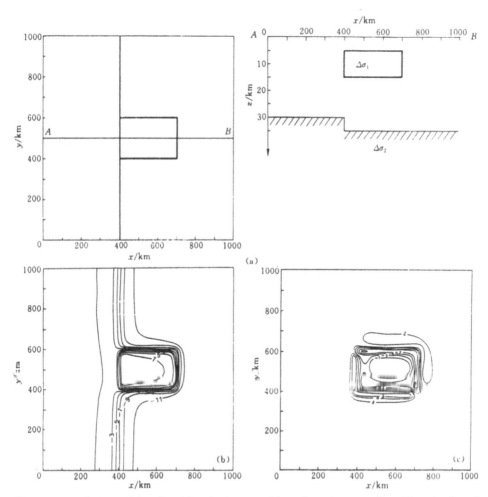

Figure 2. Synthetic model and multi-scale decomposition of gravity anomalies. (a) projection of the model in x-y plan and transverse section of the model , (b) gravity anomalies of the model , (c) difference between the first-order wavelet detail and the third-order.

On the other hand, the positive parts of the small-scale gravity anomalies shown in Figure 5 are located mainly in some of orogenic belts in China. Figure 6 shows the area occupied by the anomalies and known base-metal ore deposits. It can be clearly seen that almost all the known giant ore deposits are located in the positive part of the small-scale gravity anomalies, especially in their boundaries with lager curvature. Therefore we believe that large or giant ore deposits are tightly related to uppermost crustal high density masses. A careful study on these results shows that the high gradient zones of the small-scale anomalies are spatially coincident with base metal ore deposits, specially if the zones bend with lager curvature. These two facts, i.e., great gradient and large curvature, are selected as our criteria for geophysical prediction of base metal minerogenetic zones in China as show in Figure 7.

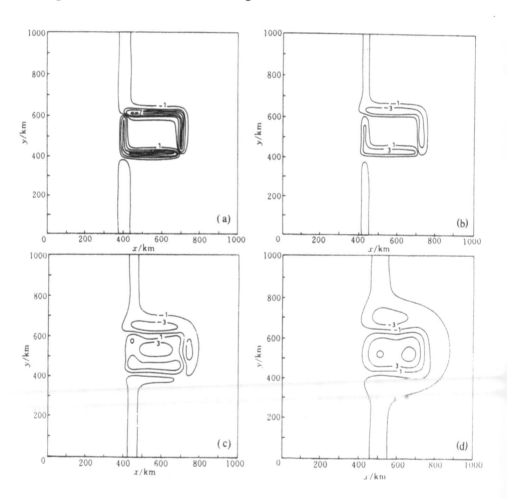

Figure 3. Wavelet details of the gravity anomalies as shown in figure 2(b). (a) the first-order, (b) the second-order, (c) the third-order, (d) the fourth-order.

The map produced in this geophysical analysis provides a new viewpoint for ore exploration in China based on the uppermost crustal density distribution and great supply of heavy masses that might inhomogeneously exist in the original crust. After Archaeozoic Era, as we understand, mineralization occurred mainly along plate boundaries by subduction, collision, or other orogenic activities. If these active zones across the original heavy-mass belts of good supply, some giant ore deposits are likely developed.

More geological interpretation of the wavelet details from the first to the fourth order has been going on. A paper about partition of Chinese crustal density structures by using the wavelet details has been published (Yang and Hou, 1996).

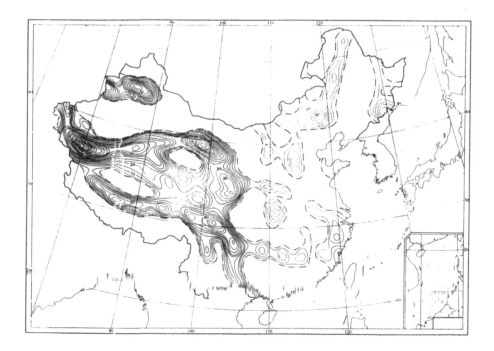

Figure 4. Contour map of the decomposed positive gravity anomalies caused by upper crust heavy masses in China.

CONCLUSION

Uppermost crustal high density zones are tightly related to the location of lager or giant ore deposits and can be used as a criterion for predicting the minerogenetic belts in exploration planning. In order to apply the criterion   new techniques for geophysical data processing should be used to decompose the potential field data into different components with different scales or sources. The two-dimensional wavelet transform and multi-scale analysis are employed for this purpose and produce desirable results for decomposition of the gravity field of China. The difference between the first and the third order wavelet details indicates the uppermost crustal density distribution, and can be used for prediction of main minerogenetic belts containing large and giant ore deposits.

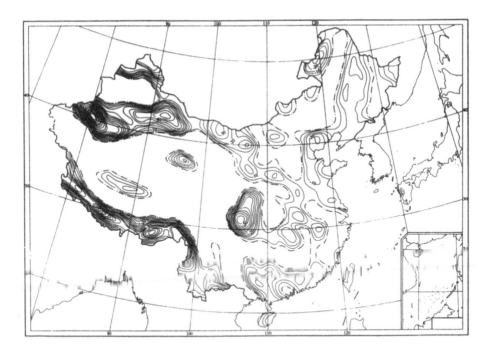

Figure 5. Contour map of the decomposed negative gravity anomalies caused by upper crust light masses in China.

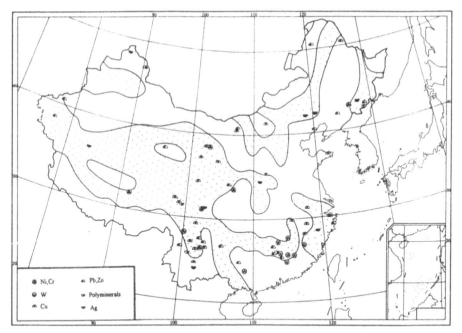

Figure 6. Comparison of the decomposed positive gravity anomalies as shown in Fig.5 with known large metal ore deposits in China.

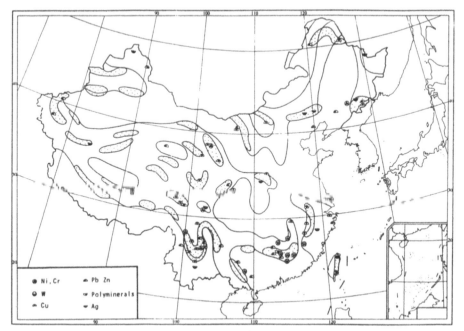

Figure 7. A prediction of main mineralization zones possible containing giant ore deposits, based on heavy mass distribution as shown in Fig.5.

References

1. Daubechies, I., Orthogonal bases of compactly supported wavelets, *Comm. Pure Applied Math.*, **41**, 909-996

   (1988).
2. Daubechies, I., *Ten Lectures on Wavelets*, Capital City Press, Verment (1992).
3, Mallat, S.G., A theory for multi-resolution signal decomposition: the wavelet representation, *IEEE, Trans.Pattern Analysis and Machine Intelligence*, **11**, 674-693 (1989).
4. Mallat, S.G., and Hwang, W.L., Singularity detection and processing with wavelets. *IEEE Trans. Infor. Theory*, **38**, 617-643 (1992).
5. Yang Wencai and Hou Zenze, Partition of Crustal Density Distribution of china, *Continental Dynamics*, **1(2)**, 32-40 (1996).
6. Hou Zunze and Yang Wencai, Two-dimensional wavelet transform and multi-scale analysis of gravity field of China, *Acta Geophysica Sinica (in Chinese)*, **40**, 85-95 (1997).

Proc. 30th Int'l Geol. Congr., Vol. 20, pp. 121-129
Liu (Ed.)
© VSP 1997

Techniques and Application of Gravity Survey with High Accuracy in People's Republic of China

ZUO YU   JIN YISHENG

The Methodological and Technical Center of Regional Gravity Survey, Ministry of Geological and Mineral Resources,710016 Xi'an , China

Abstract

Researches on high accuracy gravity survey have been fulfilled according to the System Engineering principle by the Ministry of Geology and Mineral Resources in China since late 1970s. Great progress has been made in methods and techniques. The national gravity system has been set up. Anomaly precision was improved to tens times using aerial photography and GPS techniques. we have perfected the high accuracy terrain correction with multi–sections of numeric terrain in mountainous areas, evaluation for correction errors, iterative interpretation for 3D density model with multifacets, direct processing and interpretation over a 3D curved surface, microgravity prospecting, engineering survey, and gravity & models data base. Considering "concepts" further, we discussed the colony effect of multi–density bodies, found out the constraint conditions of four types with three classes for inversion, and put forward the interpretation strategies for 3D fitting with multi–constraints and for synthetic geophysical models. The application effect on some fields are obvious. During the past ten years, we have surveyed for 6,900,000 km^{2}and submitted corresponding gravity reports and maps. New ideas have been got in crust–mantle structure, geological structure units, regional ore–controlling factors, and direct or indirect prospecting. Results in oil and gas exploration are quite successful. Microgravity has been successfully adopted to predict oil reservoir under good conditions. Good effect on probing the holes and appraising the stability on some engineering bases has also been achieved. .

Keywords: High accuracy gravity survey, Technical progress, Deepened concepts, Application effect.

INTRODUCTION

Since 1970s, research and application of the high accuracy gravity survey techniques have been made. This work was organized systematically by the geophysical branches such as the Ministry of Geology and Mineral Resources of China. Much progress has been achieved, and we have got obvious effect in the aspects of regional gravity survey, exploration for metal & petroleum resources and engineering prospecting.

PROGRESS OF METHODS AND TECHNIQUES AS WELL AS MAIN TECHNOLOGICAL RESULTS ACHIEVED

Gravimeters, techniques, and standards

We have manufactured or introduced some highly precise gravimeters, marine gravimeters and several techniques such as analytic plotting for aerial photography, GPS, computer application, and so on. We also issued seven kinds of corresponding standards, which has greatly promoted the accuracy for field gravity data acquisition, coordinate of survey station and terrain correction, density acquisition of rocks and total accuracy of the Bouguer gravity. For example, total accuracy of the Bouguer gravity in prospecting may be from 30 to 80×10^{-8}m / s^{2}or so.(Usually 30×10^{-8}m / s^{2}for smooth terrain and about 80 to 200×10^{-8}m / s^{2}in mountainous region), and the height coordinate may reach to tens centimeters. The seven technical standards we issued are as follows: Technical standard for gravity survey with the scale 1:50,000, standard for regional gravity survey with the scale 1:200,000 or 1:1,000,000, standard for compiling regional gravity maps, standard for digitizing terrain height with the net of one kilometer by one kilometer, details on aerial photography measurement for geophysical and geochemical prospecting, technical manual for measuring height by atmospheric pressure and manual of GPS measurement for geophysical and geochemical prospecting.

Gravimeter calibration sites and gravity base station networks

We have built eight national and twelve provincial sites for gravimeter calibration, and built both national and provincial gravity base station networks that consists of 24 first class provincial networks including 478 base stations, which unites nationally the instrument dimension for gravity survey, starting value for gravity survey and controlling parameters for gavity observation. The base gravity stations in China have been jointly measured with the international base station at Paris.

Data bases of gravity, gravity models, and gravity work station D

We have set up national and local gravity data base, data base for gravity models and gravity work station. The gravity data base stores all the net height data and part height data of the marine bottom over the mainland covering an area of 9,600,000 km^2 It also stores most of the gravity data from the regional gravity survey that has been accomplished recently. Besides general functions, this data base may also do the routine correction and interpretation for gravity using its sub—base consisting of programs for gravity calculation. Adopting the structure of building blocks, the gravity model data base may imitate the practical geological situation and calculate the derivatives of the models, either simple

---

D   Zuo Yu et al., Research Report on Building and Application of the Data Base for Gravity Models (No. Y85016), 1994, P.R.China

or complicated, which may be used to testify the accuracy and adaptability of the gravity data processing and interpretation methods, at the same time, it may also be used for studying the complex anomalies and improving or searching the new methods. Our gravity work station includes hundreds of optimum data processing and interpretation programs that forms the software gravity work station version 1.0 for market with bright application.

Some deepened concepts and ideas.

We put forward that the Bouguer gravity be the anomaly corrected to the observation surface and the corresponding geological–geophysical model(Fig. 1).We also put forward that forward calculation is unique, but there exist duplication or similarity among the results acquired, while the inverted results show usually constrained nonuniqueness. That nonuniqueness is because of the forward duplication and similarity. The duplication is different from similarity, for example, a sphere body and a mass point with the same mass, or a horizontal cylinder and an equivalent mass line, all have the same field respectively, while the gravity fields from some different sources are similar within the measurement precision. The nonuniqueness of the inversion is confined by four types of constraining conditions, each having three different classes. The four constraining conditions with three classes are familiar to everybody, that is to say, four types refer to known geology condition, measured density data or its general distribution range, gravity anomaly characteristics and integrated chracters of geophysical and geochemical anomalies; And the first class of the three is important and so must be followed, for example, the surface geology conditions or geology bodies encountered in drilling may be taken as starting step or controlling condition but not be contradicted with themselves. The second one is strong constraints, for instance, parameters of the density body inferred from isolated regular gravity anomalies, faults inferred from multi–anomalies, and the definitely inferred results from the seismic exploration, etc. The third or the last class is comparatively weak constraining conditions, such as the geology or physical parameters data from the neighbouring area, characters of the noisy or folded gravity, magnetic or electrical anomalies. According to the attribution of the inverse problems discussed above, we consider the integrated interpretation methods as applicable, and the methods adopt three–dimensional optimum fitting with same or different source.

It is put forward that the gravity anomalies due to relatively dense bodies have the colony effect. That is, as the density or observation distance increases, the anomalies due to each dense body will demonstrate a process from shallow interference to heavy fold and an integral whole at last. The colony effect is strongly related to the complicated anomalies and the anomaly separation. For instance, the complicated anomalies due to shallow interference may often be confused with the ones due to density body of complex shapes, while the heavily folded anomalies may be misunderstood as ones due to several smaller dense bodies at shallow depth and a bigger body at great depth, and the integral

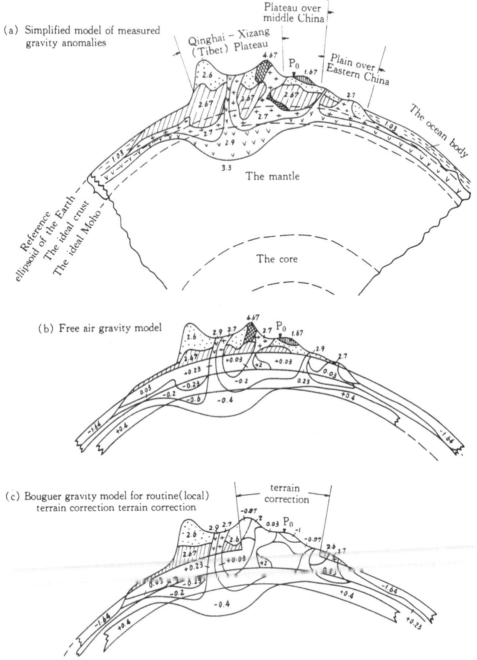

Figure 1. Sketch models of the gravity anomalies.

(a) shows the simplified model of measured gravity anomalies. (b) shows the Free air gravity model. (c) is the Bouguer gravity model for routine(local) terrain correction.

Note: the rocks with decorative pattern and number show different geologic bodies and their density values; those without decorative pattern but number are bodies after correction and their densities.

whole anomaly may only be from the main body or colony centre of gravity or equivalent density body. On the other hand, there is also the similar phenomenon when the density distribution folds vertically and lateral span of its vertical projection is small.

Technical methods suitable to the practical situation in China.
China is a big country with many mountains, it spans the latitude with fifty degrees and the longitude with sixty degrees or more. Therefore, we set up the methods for far terrain correction over the sphere surface, high accuracy terrain correction with multi–divided bodies for digital topography in mountainous area, and the evaluation strategy of the residual errors during terrain correction. The key to the high accuracy terrain correction in mountainous areas is to divide topography with multi–division and to plot the digital topography using topography map or aerial photos of the large scale, which intends to imitate the real topography most so as to let terrain correction be precise enough. The residual errors in terrain correction are evaluated by gradual line of the normalized correction values. The principle is as follows: As digitizing the topography and dividing the section bodies much finely, the curve of the terrain correction values will gradually be a fixed value which may be approximately taken as the "True Value" of the correction, the error in terrain correction can then be evaluated by the normalized correction values of high accuracy.

Appraisal of the data processing effect with the variation curve. No matter whether you use the methods of averaging within a sliding window, continuation, data smooth, data interpolation, filtering or parameter transformation, they all come to anomaly variation because of algebraic or calculus calculation. Therefore, by selecting systematically the variation parameters of the different precessing, we may get the variation curve with different variation degrees. As a result, it is possible to determine the effect of data processing and give out the qualitative or semi–quantitative controlling parameters that show the effectiveness. For example, in order to determine the regional background, some geophysicists working on gravity in China at present usually expect to weaken local anomalies by averaging within a sliding window. So the key strategy to determine the effectiveness of this method is to analyse the degree to which the local anomalies have been cut. Suppose we process an isolated local anomaly using window averaging, if the window size is so small that it only includes one observation station, obviously, the processed anomaly is the same as the original one, and if the window is infinite large, the processed anomaly will gradually tend to be zero and the covered area increases greatly. The curve between the window size and the maximum of anomaly as well as covered area may be acquired by gradually changing the window size. The curve is normalized in practical calculation(Fig. 2). When using the variation curve, you are only expected to measure the ratio of window's size to dimension of the anomaly processed. That is to say, the remained local anomalies and their co—

vered scope after window processing can thus be obtained from the curve. During practical application, the ratio of the maximum local anomaly to the local anomaly appearing most frequently must be measured to determine their variation degree by the variation curve. From the variation curve discussed above, we see that it is necessary to use and appraise multi–division, wavelet transformation or other methods.

Joint interpretation of operator with computer using two and a half dimensional models for forward and inverse problems is rather successful now. However, the geophysicist at gravity in China are now researching the three–dimensional processing and interpretation techniques. Such as, the potential transformation on a curved three–dimensional surface in which it has come true that continuation, derivative calculation and integral transformation are done from curved to curved surface as well as from curved to flat surface for irregular net stations on the microcomputer; Potential field calculation for arbitrary bodies of polyfacet; Joint inversion of operator with computer for 3–D models; Improved forward and inversion algorithms for single and multi–layered density interfaces, and so on.

Two methods are used for modelling variable density distribution at present. One is the unit element method, another is to calculate potential field by taking the density distribution as a function of depth, usually being linear or exponential. In reality, there exists the key problem that we are not able to know exactly the distribution law of the rocks beforehand.

As to computer plot for the interpreted results, our country may plot the colored plane map, section map and three–dimensional map based on the geological–geophysical data by means of the microcomputer.

APPLICATION EFFECT OF HIGH ACCURACY GRAVITY SURVEY IN SOME FIELDS IS SATISFACTORY

New gravity investigation with various scales

During the past ten years, we have systematically done the new gravity investigation with various scales for about 6,900,000 km^2 and submitted corresponding national, regional and hopeful ore–controlling gravity maps as well as reports, which include the Bouguer gravity map in continental China with the scales 1:5,000,000 and 1:4,000,000 as well as 1:2,500,000 that is going to be completed, the Bouguer gravity map of eastern China with the scale 1:1,000,000, the Bouguer gravity map over the marine area of southern China and the neighbouring district, the gravity map of 1:500,000 on the six important ore–controlling belts in China that are in northern boundary on Huabei Platform, Qinling–Dabashan district, district of southern China, middle and lower part of the Youngtze River, district surrounding Yunnan Province and Tibet Autonomous Region and Tancheng–Lujiang district, and the one hundred

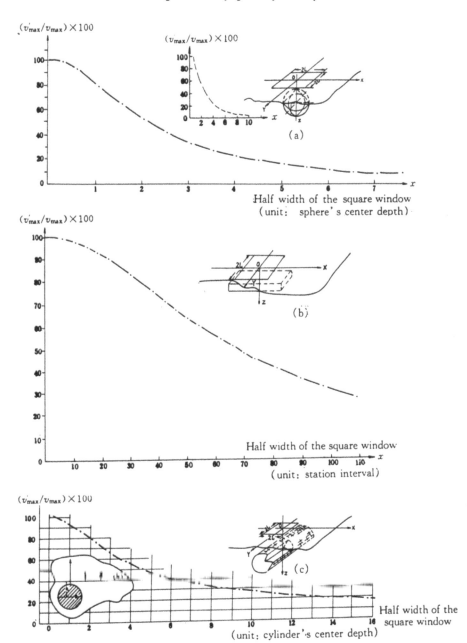

Figure 2. Percentage variation curve of the maximum values from sliding window processing. Models used include 2, 2.5, and 3 - D ones: (a) 3 - D model, the sphere with center depth h. (b) 2.5 - D model, the rectangular prism with dimension $40 \times 10 \times 5$ and up surface depth of 10 station intervals. (c) 2 - D model, a cylinder with center depth h.

kilometer wide gravity corridor map over the GGT geosections that are from Quanzhou in Fujian Province to Keketuohai in Xinjiang Autonomous Region, from Luoyang in Henan Province to Zigui in Hubei Pronvince, from Yiexian in Henan Province to Nanzhang in Hubei Province, from Lingbi in Anhui Province to Beiyunebe in Inermongolia Autonomous Region, from Chanbeishan mountains to Manzhouli in Northeastern China, and so on. These maps demonstrate the gravity characteristics of the modern crust—mantle structures and young evolved continental platforms of the multi—basin orogenic belt system in China. For instance, the gravity gradient belts, the Daxinganling—Taihangshan—Wulingshan Belt and the belt surrounding Qinghai—Tibet Plateau, are due to the two gradient belts of the Moho discontinuity of our country. And the nuclearpart of the gravity high and magnetic high as well as the surrounding low gravity belt on every big basin in China describe basin's formation during the latest orogenic period, at the same time, we may also see the gravity anomaly characters of the earlier orogenic belt and the solidification erosion in large basins. In some important districts with hopeful ore—controlling index, we have done the gravity survey with scale 1:200,000, which play a great role in dividing faults, locating complex distribution and determining the sub—structures in geotectonic units. For example, on the gravity map of 1:200,000, the Daxinganling gradient belt may be divided into several smaller gradient belts, which indicates that the gradient belt is a stairs—like one. Another example shows many big gravity lows exist in East Qinling district, which indicates that a great deal of granite complex with crust origin locate here.

Gravity investigation in metal ore districts
Some gravity investigation has been done with scale 1:10,000 or 1:50,000 in the metal ore districts, the accuracy is generally about 50 to 200 × 10^{-8}m / s^{2}microgals. These data have played an important part in re—recognizing geological structures, ore—controlling factors and prospecting for high(low) density mineral resources in the district. For example, in the basin of igneous type at low part of the Youngtze River, the igneous rock had originally been thought as rather thick. Instead, it was proved to be thin by the gravity, like a quilt covering the sediments, which enlarge the application of prospecting sedimentary ores over the covered area. Furthermore, the indirect ore—prospecting with gravity was also successful in Hunan Province, the metal ores here are located nearby the actilite, therefore, the ores may possiblly be picked out surrounding the granite complex that are recognized by gravity low. We have also some examples of direct ore—prospecting of the rich iron by gravity high and searching low density salt by gravity low.

High accuracy gravity survey in oil exploration
We have applied the high accuracy gravity to oil exploration for ten years in some oilfields or in newly opened districts in China. Since gravity accuracy has been improved tens times, the gravity survey becomes more powerful in locating

traps and recognizing deep structure information. The best effect is achieved where high quality data cannot be acquired by seismics, mountains exist, and in the districts of belts between tides, shallow marine with less than five meter water in depth and thickly inhabited place with developed industry or agriculture. If the terrain is flat and there are good conditions for oil developing and storage, gravity may be applied directly to search for oil & gas traps and point out the drill position, as a result, some of them have been drilled to be industrial oil. At present, experiment is also being done to predict oil & gas distribution by detecting the lost anomalies due to trap structure using microgravity measurement. Therefore, high accuracy gravity has become one main geophysical method besides the seismics on oilfield. If the density contrast is satisfactory enough, the local structure with about three thousand meters and density interface with four to five thousand meters in depth may be recognized. For instance, a trap in depth of 3,000 meters has been recognized by high accuracy gravity data and then been drilled to be industrial oil on an oilfield in eastern China.

High accuracy gravity survery in engineering and archaeology
The high accuracy gravity survey has been put into use in the engineering and archaeology, such as probing natural or artificial holes and appraising the stability of some large engineering bases.

REFERENCES

1. Jin Yisheng,Liao Yongwei and Li Zhi et al., *Specification for regional gravity survey*, Beijing: Geological Publishing House, China(1983).

2. Xiao Jingyong,Zhou Guofan and Liu Zhenmin et al., *Handbook of data interpretation for gravity exploration*, Beijing: Geological Publishing House, China(1983).

3. Pan Zuoshu, *Data processing and interpretation for gravimetric data*, Beijing: Geological Publishig House, China(1992).

Proc. 30th Int'l Geol. Congr., Vol. 20, pp. 130-139
Liu (Ed.)
© VSP 1997

Use of Induced Polarization Method in Search for Deposits of Composite Geological Media

G.N. Shkabarnya, N.G. Shkabarnya, B.L. Stolov
Department of Geophysics, Far Eastern State Technical University, Vladivostok, Russia

Abstract

Various modifications of induced polarization method (IP) are widely applied in Russia to attain some important geological objectives, first and foremost in exploration for ore mineral and underground waters. Specifically, in Primorye, the induced polarization vertical electric sounding method (IP-VES) is practically unique in search for hidden deposits of complex ores, tin, wolfram and copper enabling to perform deep mapping of a section and to reveal zones of disseminated and veinlet-disseminatied sulphide mineralizations. Investigations by means of this method can be conducted in areas of composite geological structure having within geoelectrical section and under the condition of heavily rugged topography various subvertical and tilted heterogeneities and local objects. Local structures of ore control and ore localization are characterized, as a rule, by low resistivity and high polarizability. Among these are areas of intensive sulphide mineralization associating with zones of abnormally high fissibility. So, in Dalnegorsk and Kavalerovo ore districts all deposits and ore occurrences are within the limits of the IP anomalous areas. With the help of this method some promising deposits of underground waters, including Pushkinskoye deposit, have been prospected.

Specific conditions of application of IP method have required special approach to the field investigations technique as well as the interpretation technique for the material obtained. For this purpose various modifications of the field technique using IP-VES in complicated conditions of mountains and taiga in new areas as well as in ore field flanks have been developed. In this case the depth of exploration from hundreds of meters up to one and a half kilometer can well be achieved. But the successful use of IP method largely depends on the possibility to develop a close and authentic physico-geological model enabling to determine reliable parameters of the objects under investigation. Experts are familiar with automatic systems of electric sounding interpretation in which data of resistivity method receive primary emphasis. The suggested technology of IP interpretation is based on mathematical modeling of horizontally-heterogeneous media, satisfactorily approximating real geoelectrical sections and representing combinations of smooth, tilted and sharply sloping layers including local heterogeneities. Preliminary analysis of typical geoelectrical sections of ore districts has been conducted, algorithms and programs for a number of horizontally-heterogeneous media models have been developed, peculiarities of spatial structure of IP for some models and configurations have been studied, algorithms of preliminary data process and initial approximation of models have been developed.

Keywords: IP method, geoelectrical model, heterogeneous media, modeling, interpretation, approximation.

INTRODUCTION

To attain some important geological objectives, first and foremost in exploration for ore mineral and underground waters various modifications of the induced polarization method (IP) are widely applied in the Far Eastern region of Russia. Specifically, in Primorye, the induced polarization vertical electric sounding method (IP-VES) is practically unique in search for hidden deposits of complex ores, tin, wolfram and copper enabling to perform deep mapping of a section and to reveal zones of disseminated and veinlet-disseminatied sulfide mineralizations.

Investigations by means of this method can be conducted in areas of composite geological structure having within geoelectrical section and under the condition of heavily rugged topography various subvertical and tilted heterogeneities and local objects. Local structures of ore control and ore localization are characterized, as a rule, by low resistivity and high polarizability. Among these are areas of intensive sulfide mineralization associating with zones of abnormally high feasibility. So, in Dalnegorsk and Kavalerovo ore districts all deposits and ore occurrences are within the limits of the IP anomalous areas.

In search for underground waters the layered geoelectrical sections under investigation also contain local inclusions, gently sloping and inclined pinching out horizons, and often even steep inhomogeneities. The water horizons, depending on conditions, are characterized by low or increased resistivity values in comparison with enclosing rocks, but their polarizability values are more often lower. With the help of the IP method some promising deposits of underground waters, including Pushkinskoye deposit, have been prospected in Primorye.

The marked features of the ore districts and hydro-geological provinces in other regions of the Far East and Russia [1] are similar to those of Primorye, therefore they can be considered typical for the consideration of the methodical features and opportunities of the IP method in search for mineral deposits of composite geological media. Specific conditions of application of IP method have required special approach to the field investigation techniques and to the interpretation technique for the material obtained.

FIELD WORK TECHNIQUES

The researches by the IP method were conducted, basically, at the search stage on a scale of 1:10,000 - 1:25,000, and in deep searches and in the study of promising territories - on a scale of 1:50,000 - 1:100,000. During the work use was made of the developed electrical profiling (EP IP) and sounding (VES-IP) arrays. The main electrical profiling arrays used were symmetric four-electrode ($AMNB$), three-electrode (AMN), combined (direct AMN and return MNB), axial dipole and middle gradient (MG IP). Usually the size of the AB current lines did not exceed 200 m, except for the MG IP modification, where the AB size reached 2000 - 4000 m; the size of the MN receiving lines was 50-100 m, and the step of moving MN equaled 50 m. In sounding use was made of symmetric four-electrode, combined three-electrode, Wenner ($AM=MN=NB$), axial ($AB=BM=MN$) arrays and a system of three-electrode arrays with the fixed current line and mobile MN receiving electrodes. It has been established, that the most effective for the study of the sections under research should be considered combined arrays and fixed current line arrays.

Apparent resistivity (ρ_a) and apparent polarizability (η_a) were determined by direct and alternating currents. In the former case long charge modes (1-3 minute) were used, and in the latter case low frequencies (0,62 Hz) of alternating current were applied, phase shift in degrees (φ_{ip},°) being originally determined. The alternating current would be appropriate for use in terms of malfunctions.

GENERALIZED GEOELECTRICAL MODEL OF THE ORE DISTRICTS

To get the accurate interpretation of the field materials obtained, the geological structure of each of ore districts were given a detailed study. Under investigation were rocks and their physical properties, tectonic peculiarities and mineralization distribution patterns. The analysis of the work done revealed the main features of the geoelectrical models in most of the ore districts and showed that the districts have a number of geologic features in common. The conceptual model (Fig. 1) has three clear cut generalized structure-material complexes, which differ in thickness and the completeness of exploration, and vary in differentiation of resistivity and polarizability.

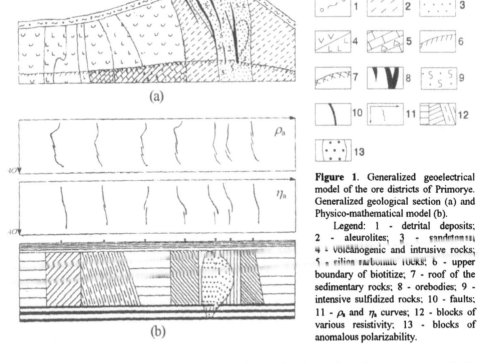

(a)

(b)

Figure 1. Generalized geoelectrical model of the ore districts of Primorye. Generalized geological section (a) and Physico-mathematical model (b).

Legend: 1 - detrital deposits; 2 - aleurolites; 3 - sandstone; 4 - volcanogenic and intrusive rocks; 5 - silicic carbonate rocks; 6 - upper boundary of biotitize; 7 - roof of the sedimentary rocks; 8 - orebodies; 9 - intensive sulfidized rocks; 10 - faults; 11 - ρ_a and η_a curves; 12 - blocks of various resistivity; 13 - blocks of anomalous polarizability.

The top complex (detrital deposits) is most thoroughly investigated and represented, in the majority of cases, by gently sloping layers showing sharp variability of electrical properties (ρ_1, η_1) and characterized as the products of decomposition, oxidation and freezing of rocks. The values of thickness average 30-50 m, but sometimes reach 100-150 m. This complex is not of much interest as far as possible zones of mineralization are concerned, but its structure and geoelectrical parameters should be given a detailed study in order to avoid errors in the interpretation of the fragments of electrical sounding curves, representing underlying structural stages.

The middle complex, 300-1000 m in thickness, is a set of individual blocks among which we observe the predominance of steep horizons caused by tectono-magmatism activity; or gently sloping and inclined layers, resulted from sedimentogenesis, or a combination of both. Among the terrigenous and volcanogenic sediments the main objects of investigation are sulfidization areas, various folds, ruptures and grinding zones. Resistivity of rocks (ρ_2) ranges from 20 to 10000 Ωm and more, and their polarizability (η_2) - from 2 to 30 per cent. Of special interest in the search work are local objects of various stretches, as well as fracturing and grinding zones showing high values of conductivity and polarizability.

The lower complex is made up of dense terrigenous sediments, granitoid massifs and zones of hornstone formation, including regional biotite metasomatites. The rocks are folded and form either anticline or syncline structures, their upper boundary being gentle or gently inclined. The resistivities of the sediments are high ($\rho_3 > 10000$ Ωm), and the polarizabilities are low ($\eta_3 \approx 1 \div 3$ per cent).

It has been established that the electrical resistivities of the structure-material complexes vary over a wide range, but regarding polarizability they are less differentiated. The radical difference from the enclosing rocks has been noted only in sulfidized or graphitized rocks. The analysis of data obtained from the modeling of heterogeneous media polarizability fields let us establish that variable inhomogeneities of the reduced polarizability ($\eta_2/\eta_1 < 1$) and increased resistivity ($\rho_2/\rho_1 > 1$) are not practically represented by the IP curves (see the fig. 1).

INTERPRETATION OF THE FIELD MATERIALS

The interpretation of the electrical profiling data was at the first stage restricted to the qualitative identification of anomalous areas and localization of the boundaries of the section using the average maximum of the anomalies [1, 2]. The apparent polarizability isoline maps gave an idea of the elements of the geological structure, including tectonic disturbances, zones of sulfide mineralization and unchanged rocks. At the second stage use was made of quantitative methods based on the analysis of the data obtained from the mathematical modeling of horizontally-heterogeneous media IP fields [3].

The selected horizontally-heterogeneous media (Fig. 2), in a first approximation reflect the main objects of conceptual model and has a mathematical solution for determining point source field potential. They include semispaces having horizontal or vertical or inclined boundaries and also involving wedge-shaped or local objects. To fit them the sophisticated mathematical boundary-value problem techniques [5] were simplified through the methods of separation of variables and integral equations to the expressions of potential calculating [4, 3]. Based on the solutions, provision is made of a software providing a means of modeling IP fields to study their spatial structures in different parameters of medium and arbitrary arrays and to use the selection technique in the interpretation of the field IP data.

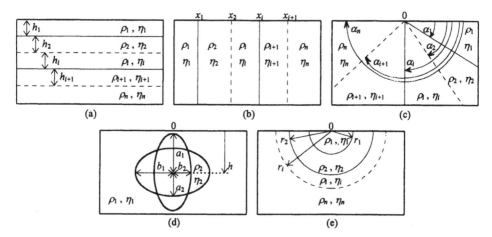

Figure 2. Elementary physico-mathematical models of horizontally-heterogeneous media. (a) - model of horizontally-stratified medium, (b) - model of vertically-stratified medium, (c) - model of wedge-shaped medium, (d) - model of medium with spheroid, (e) - model of medium with hemispherical boundaries.

The calculation of the apparent polarizabilities being possibile, the accuracy of the construction of geoelectrical sections increased dramatically, particularly with the use of close-spaced sounding points. The data are processed on a computer in several steps. Each step involves certain operations, beginning with preparation of data, computational procedures and ending with the graphic data presentation.

The initial stage provides the preliminary processing and the construction of the initial approximation. It involves visualization of the sounding curves, construction of the sections and maps by the initial data and various transformations, division of the area under research into districts of similar models of medium, calculation and construction of theoretical curves for different media, processing of parametric curves with further reference of geoelectrical boundaries to geological ones, and estimation of the effect of topography and surface inhomogeneities. In practice this stage often provides the solution of the problem of medium investigation.

In promising areas quantitative interpretation estimating geometrical and physical parameters is used. Here the selection methods are used in dialogue conditions. As a result both the model itself and its parameters are refined.

An example of the approximation method of IP-VES data interpretation is the Proninsky area (Fig. 3), located in the Dalnegorsk ore districts. Here using the data of the VES-IP method, gravity surveying and geochemical prospecting data as a base an anomalous area has been preliminaryly revealed, the position of the roof of the

APPROXIMATION MODELLING OF ORE REGION'S GEOELECTRICAL SECTION BY IP-VES DATA

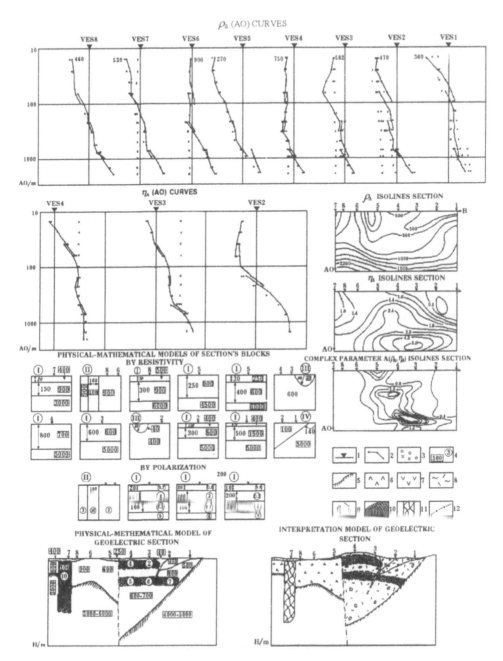

Figure 3. Proninsky area. Legend: 1 - IP-VES points; 2 - initial ρ_k and η_k curves; 3 - theoretical ρ_k and η_k curves; 4 - Ωm resistivity and percentage polarizability values; 5 - marker geoelectrical horizon; 6 - dacites, dacite-andesites; 7 - andesites; 8 - ignimbrites of rhyolites, tuffolava; 9 - agglomerate tuffs; 10 - zones of sulfide mineralization by IP-VES data; 11 - sulfidized fractures by IP-VES data; 12 - fault.

underlying high-ohm silica-carbonate rocks has been located. At the interface of these rocks and the overlying volcanics, the existence of the sulfidized complex mineralization sediment lens was expected.

On the calibration profile the apparent polarizability curves are characterized by a smooth shape and appear as two- or four-layer. The apparent resistivity curves are distorted by fractures and show regular apparent resistivity increase as the AO spacing increases. In the curves a considerable discrepancy of values of apparent resistivities and polarizabilities is observed in going from the smaller receiving line to the larger one. From the detailed analysis of the curves and qualitative sections an initial approximation of the model involving from thin drift, fragments of laminated medium, steep inhomogeneities and an inclined boundary has been obtained. In the vicinity of the points 3-4 (Fig. 3) there are two anomalous objects arranged above the other showing increased polarizability. Throughout the profile, in the section there exists a high-Ω index horizon, whose dipping varies from minimum at the ends of the profile to maximum in the center.

To refine the model and its parameters quantitative interpretation of the curves was performed. For this purpose hemispace models of vertical layer, horizontal boundaries, inclined contact and subsurface hemisphere were used (Fig 3). In doing so, during the correlation of the practical curve to the theoretical ones not the entire function but only its separate fragments were involved. For example, the initial parts of the curve could be correlated with the "inclined contact" model, and the end parts of the curve – with the model of laminated medium. The correlation of all the fragments of the model resulted in the construction of a physico-mathematical model, and after giving it geological content – the final geoelectrical calibration profile section (Fig. 3).

Another example of the approximation method of VES-IP data interpretation is the Avgustovsky area (Fig. 4, 5), located in the vicinity of the Smirnovskoye tin complex deposit. The area is characterized by a complex geological section made up of the blocks of horizontally-heterogeneous media and having heavily rugged topography. The most favorable for ore localization here are fault zones, within the limits of which at frameworks of the investigated area are located large ore bodies of Smirnovskoye, Vstrechnoye, Yuzhnoye deposits. These zones are noted by decrease of resistivity from 200 Ωm against the background of 1000-3000 Ωm and increase of polarizability up to 10 % among 1-2 % host rocks. The work has been conducted with the application of closely spaced array systems on the network 500×100 m using three-electrode arrays and maximum spacing 2000 m.

Based on the visual study of the sounding curves, initial data sections and various transformation sections it has been established that in the medium of the middle complex under investigation there are extended steep horizons. There are also inclusions of abnormally high polarizable and conductive local objects. The lower portion of the section is represented by the laminated high resistivity and low polarizability rocks.

With the purpose of further approximation interpretation by means of mathematical modeling the effect of the topography has been estimated [3]. The calculated theoretical model curves, which approximate the topography, have shown that for separate points the distortions reach 40 %. The distorting effect of the relief has been reflected in a number of corrections of the initial sounding curves in a certain portion of the area.

The preliminary interpretation of the curves showed that the section cannot be approximated by any known horizontally-heterogeneous physico-mathematical models. In this connection, as it was shown by the previous example, it must be divided into separate blocks, within the limits of which certain objects dominate. For the main models we applied "ridges" and "valleys" with infinite slopes and boundaries, reaching the surface along the rib line; vertically-stratified and horizontally-stratified sections (Fig. 4).

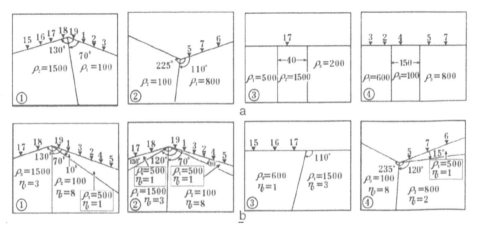

Figure 4. Elementary horizontally-heterogeneous physico-mathematical models for apparent resistivity (a) and apparent polarizability (b) curves interpretation.

The quantitative interpretation of fragments using several models was conducted in the interactive mode [3]. In each of the blocks a wide range of physical and geometric parameters was set which then decreased in each further approximation. In each of the approximations the number of the boundaries often decreased. During the selection the practical curves were made satisfactorily coincided with the same model, or the joint influence of two or three models was taken into account. The thickness and the electrical properties of the layers of the top part of the section were defined by the familiar methods, used for layered media.

As a result of the interpretation some parts of the section structure and physical properties of rocks were specified, which was reflected in the physico-mathematical model (Fig. 5, a). Resembling blocks have been collected into the geoelectrical model along the profiles. Having located the geoelectrical boundaries with the geological ones geologo-geophysical sections were built (Fig. 5, b). On the basis of these sections locations of wells were recommended within the limits of the zones of sulfidized fractures containing ore bodies.

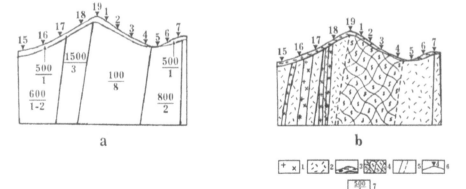

Figure 5. Avgustovsky area. Physico-mathematical model of a section (a) and geologo-geophysical section (b) along one of the profiles. Legend: 1 - dikes; 2 - volcanogenic sediments; 3 - surface layer of decomposed rocks (detrital deposits); 4 - block of sulfidized fractures; 5 - faults; 6 - VES-IP points on profile; 7 - Ωm resistivity and percentage polarizability values.

CONCLUSIONS

The data provided by the use of the IP methods represent useful information on the structure of ore areas and, primarily, on the distribution of sulfide mineralization zones and ore control structures into depth. High effectiveness of the sounding modification IP method used in search for the above mentioned zones resulted from the decrease of resistivities and increase of polarizabilities. The analysis of the modeling data showed that the apparent polarizability curves reliably indicate objects showing this type of ratio of parameters and do not register host rocks of low conductivity and polarizability.

For the study of composite geological media closely spaced array systems should be used. On the basis of the analysis of the data obtained from modeling electric fields in horizontally-heterogeneous media it was established that the most effective arrays for closely spaced systems are three-electrode combinations with highly movable sounding center and fixed transmitting lines with highly movable receiving electrodes. The step of the sounding center movement and receiving line movement in the array system is defined by the concrete geological problems.

On the basis of systematization of electric properties of rocks and ores, and generalization of the investigations conducted with the IP method a conceptual model of some basic ore areas in Primorye has been developed. This model includes: detrital deposits sharply differentiated in physical properties; anomalous highly conductive and polarized local inhomogeneities among vertical and inclined layers within heterogeneous volcanogenic-sedimentary formation; high-resistivity and low-polarizability index horizon. The peculiarities of the generalized sections provided the basis for the choice of physico-mathematical models and the development of interpretation technique.

The construction of the geoelectrical section by the field data became possible owing to the solution of the direct problems, the development of the algorithms and apparent polarizability computer programs for physico-mathematical models which quite satisfactorily approximate real standard media and can be applied for manufacturing arrays. The analysis of the peculiarities of the electric field spatial structure by the models and arrays, the choice of the criteria of detecting in a section boundaries and objects of a certain type, the development of the methods of preliminary and quantitative estimation of the parameters of a medium provided the basis for the interpretation system.

Interpretation process consists of three - four interrelated stages. In the first stage visual survey and transformation of the initial data is conducted which result in forming a general notion of the nature of the section. During the second stage the initial approximation model is constructed. This stage includes the division of the section into similar blocks and the estimation of the physico-geometric parameters within each of the blocks. In the third stage interactive or automatic choice of quantitative parameters using elemental fragment-block approximation models is completed. In closing the fragments of the section are correlated with the general geoelectrical model and the latter is given geological content.

The outlined suggestions on the methods of doing field work, on the features of a geoelectrical section and, primarily, on the interpretation of the field materials provided an essential increase of the efficiency of the IP methods in the study of sections and in search for mineral deposits. By now the great amount of experience has been accumulated on the use of these methods on vast areas of the Primorski Krai of Russia.

REFERENCES

1. V.A. Komarov. *Electrical Prospecting by the Induced Polarization Method* Leningrad : Nedra Publishing House (1980), (in Russian).
2. B.K. Matveyev. *Electrical Prospecting*. Moscow : Nedra Publishing House (1990), (in Russian).
3. N.G. Shkabarnya. Automated interpretation of VES and IP-VES curves in ore districts, *Exploration and mineral wealth protection* 11, 40-45 (1986), (in Russian).
4. N.G. Shkabarnya. Electric field above inclined-layered medium and simple geometrical shape inclusions. In: *Electrical Prospecting*. Reference book for geophysicists. volume 1, pp. 52-59. Moscow : Nedra Publishing House (1989), (in Russian).
5. M.S. Zhdanov. *Electrical Prospecting*. Moscow : Nedra Publishing House (1986), (in Russian).

Proc. 30th Int'l Geol. Congr., Vol. 20, pp. 140-147
Liu (Ed.)
© VSP 1997

High Resolution Curie Depth and Stratum Interface Inversion Using Aeromagnetic Anomaly

ZHANG MINGHUA

Institute of Resource Engineering, Beijing University of Science and Technology, Beijing ,100083,China

GUAN ZHINING

Department of Geophysics,China University of Geosciences, Beijing,100083, China

Abstract

Based on a study and analysis of the conventional inversion methods of potential field, we use a magnetization function to fit magnetization distribution in the crust and develop a new forward-inversion combined method for magnetic strata interfaces and Curie point depth determination calculation. This method overcomes disadvantages of conventional methods, gives out much higher resolution, and it is timesaving as well. We put the method into application in Chinese national "Climb-B" geological research project, which is aimed for large and super-large ore deposits prospecting. valuable inversion results are obtained both for further geological research work and for prospecting nonferrous ore deposits in north Huabei platform, China.

Key words: Magnetization function, High resolution, Stratum interface, Magnetic anomaly.

Magnetization Function and Magnetic Stratum Model

About the source and distribution of regional magnetic field and how it is developed, different researchers have different opinions. So far, what we know are almost from the studies of the lithosphere and continental scientific drilling. Based on physical property

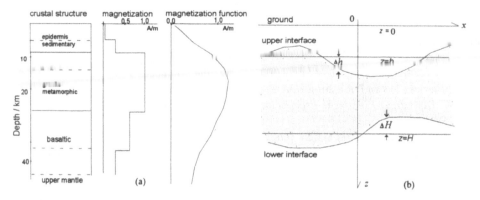

Figure 1. Illustrations of magnetization function and magnetic stratum model. (a)Magnetization variation in the crust and the magnetization function. (b)Magnetic stratum/layer model.

profiles of different rocks, strata in the crust and Curie temperature points of magnetic minerals[1,2,3,6,8], we suggest that magnetization of rocks is different at different depth in the crust and, below Curie point depth(or simply say Curie depth) where temperature is higher than Curie points of magnetic minerals, the rocks become paramagnetic. Magnetization of rocks is also different in different geological structure units. This is the main magnetic character of the crust. We thus put up a magnetization function which can match the magnetization variation in the crust much better, see figure 1(a). The function is as follows,

$$J(\xi, \eta, \zeta) = a(\xi, \eta)\zeta^n \exp[-b(\xi, \eta)\zeta] + c(\xi, \eta) \tag{1}$$

where, $J(\xi, \eta, \zeta)$ is magnetization function. $n=1$, or 2. $a(\xi, \eta)$, $b(\xi, \eta)>0$, $c(\xi, \eta)\geq 0$ are variables. They are vary horizontally with different magnetic, or geological units. ζ is a variable in vertical direction, or varies with depth.

The magnetic strata model we suggested for regional magnetic anomaly is illustrated in figure 1(b). Both upper and lower interfaces are rolling. The lower interface is at Curie depth. Above the upper interface, weak magnetic or nonmagnetic strata are supposed to exist. Below the lower interface, i.e. Curie interface, rocks become paramagnetic. Between the two interfaces, magnetization distributions, both in horizontal and in vertical direction , are as what we suggested above.

Forward and Inversion

In order to interpret deep geological structures and to learn depth variations of Curie interface and upper interface of Archean group using aeromagnetic anomalies in China national geological research project "Climb-B", we studied all the conventional methods and found none of them is suitable for getting high resolution result, either because the magnetic parameters they used are unable to represent the character of the crust, or because there is too much filtering to anomaly data, or to depth data during inversion[2,9,4]. Based on the magnetization function expression (1) and the above source model for regional magnetic field, we developed a forward-inversion combined iteration method for magnetic interface depth inversion. Let's call it MIDI. We give here, for saving page, the calculation formulas in the case of $n=1$ and $c(\xi, \eta)=0$.

Fourier transform of vertical component $Z(x,y)$ after reduction-to-the-pole is expressed as follows[2] ,

$$Z(x,y)^{uv} = \frac{\mu_0}{4\pi} (2\pi f)^2 F\{ \int\int \int_{-\infty}^{H(\xi,\eta)} \int_{h(\xi,\eta)} J(\xi, \eta, \zeta) R(x-\xi, y-\eta, \zeta) d\zeta \} d\xi d\eta. \tag{2}$$

where, $Z( )^{uv}$ and $F\{ \}$ represent Fourier transform or spectrum. $J(\xi, \eta, \zeta)$ is magnetization. $h(\xi, \eta)=h+\Delta h(\xi, \eta)$ and $H(\xi, \eta)= H+\Delta H(\xi, \eta)$ are the upper and lower rolling interface depths respectively, $\Delta h(\xi, \eta)$ and $\Delta H(\xi, \eta)$ are relief intensities, h and H are average depths. $R(x-\xi, y-\eta, \zeta) = [(x-\xi)^2 + (y-\eta)^2 + \zeta^2]^{-1/2}$. $F\{R\} = \exp(-2\pi f) \exp[-2\pi j(u\xi + v\eta)] /f$.

With the utilization of Talor series expression,

$$\exp(b-2\pi f)\Delta h = \sum_{n=0}^{\infty} \frac{(b-2\pi f)^n}{n!} \Delta h, \tag{3}$$

and put formula (1) into (2), we finally come to the forward formula, the magnetic field spectrum expression of the rolling strata model with variable magnetization distribution,

$$Z(x,y)^{uv} = \pi f \mu_0 \left(\frac{1}{b-2\pi f} \{ H \exp[(b-2\pi f)H] - h \exp[(b-2\pi f)h] \} \, a(\xi,\eta)^{uv} \right.$$

$$+ \frac{1}{b-2\pi f} \{ \exp[(b-2\pi f)H](H(b-2\pi f)+1) -1\}\{ \, a(\xi,\eta)\Delta H(\xi,\eta) \}^{uv}$$

$$+ \frac{1}{b-2\pi f} \{ \exp[(b-2\pi f)h](h(b-2\pi f)+1) -1\}\{ \, a(\xi,\eta)\Delta h(\xi,\eta) \}^{uv}$$

$$+ \frac{1}{(b-2\pi f)^2} \{ H(b-2\pi f)\exp[(b-2\pi f)H]-1\}\sum_{n=2}^{\infty} \frac{(b-2\pi f)^n}{n!} \{a(\xi,\eta)\Delta H^{\,n}(\xi,\eta) \}^{uv}$$

$$+ \frac{1}{(b-2\pi f)^2} \{ h(b-2\pi f)\exp[(b-2\pi f)h]-1\} \sum_{n=2}^{\infty} \frac{(b-2\pi f)^n}{n!} \{a(\xi,\eta)\Delta h^{\,n}(\xi,\eta) \}^{uv}$$

$$+ \frac{1}{b-2\pi f} \exp[(b-2\pi f)H] \sum_{n=1}^{\infty} \frac{(b-2\pi f)^n}{n!} \{ \, a(\xi,\eta)\Delta H^{\,n+1}(\xi,\eta) \}^{uv}$$

$$+ \frac{1}{b-2\pi f}\exp[(b-2\pi f)h] \sum_{n=1}^{\infty} \frac{(b-2\pi f)^n}{n!} \{a(\xi,\eta)\Delta h^{\,n+1}(\xi,\eta)\}^{uv}). \qquad (4)$$

Vertical component $Z(x,y)$ can be obtained either by converting from total anomaly $\Delta T(x,y)$ or from direct observations.

From the forward formula we can see, if any one of the two interface depths is given, the other interface depth can be determined. Counting out the series part of the formula, we can use the remaining part for direct inversion. This part is thus called direct inversion formula. Combining the direct inversion formula together with the series part, we can easily build an iteration procedure to calculate the interface depth values. When calculated depth difference or spectrum value difference between two consecutive iterations is smaller than a given accuracy limit, inversion or iteration stops.

Method Performance

In the case of a constant magnetization, method MIDI gives out the same results as commonly used conventional methods. While magnetization is not a constant, theoretical tests on different models have shown that MIDI is always convergent to the model depth, no matter how big the relief intensities are and, even if depth variation is discontinuous. (This is practically useful in area where large Faults develop.) And, it is comparatively time saving. Figure 2(a) is an example. Accuracy limit is 5%. Iteration

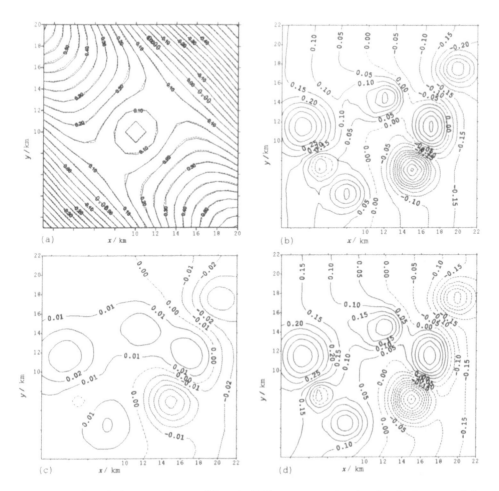

Figure 2. A theoretical test on a given model (a), (solid line represents given interface depth contour, dsah line is inversion result.) and a comparison of our method with Parker's method, (b) is interface depth model , (c) is inversion result contour of parker's, (d) is our inversion result within the same computing time period as Parker's. Unit of contour data is km.

times is 355. Time consumption on a 4M ILAM 486 computer is 77 minutes. We can clearly see that inversion depths (dash lines) are almost the same as the given model depths (solid lines).

One character of method MIDI is that it requires no filtering process to any of the data, neither to the field data, nor to the depth data before, in or after inversion. We have made a comparison with Parker's method[4,10], see figure 2(b),(c),(d). Figure 2(b) is the rolling depth interface model. (c) shows the inversion result of Parker's. (d) is the inversion result of MIDI. Computing time used by MIDI is limited to be the same as that of Parker's, 20 minutes. It is obvious that MIDI has a much higher depth resolution.

A Discussion on Average Depth

It is difficult to determine average depth in potential inversion. From the method we put up here, it can be seen that owing to the strong constraint of magnetization function in inversion, the average depth of an interface can be calculated, or adjusted. Figure 3 is an example showing this. The model we used is as in figure 2(b). The real average depth of upper interface is 2.0 km, lower interface is 5.0 km. We calculate the field data

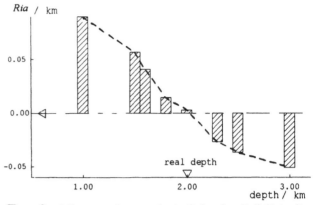

Figure 3. Adjustment of average depth of a interface. Relief intensity average value is zero at real depth of the interface. The value is not exactly zero due to computation error.

through forward procedure, then we use the data for upper interface inversion. We suppose that the average depth of upper interface is unknown. We choose several different values for this depth in inversion. We found that if the value is bigger than 2.0, the relief intensity average(Ria) is negative,   if the value is smaller than 2.0, the Ria value is positive, the bigger the value is, the smaller the Ria is. At the value 2.0, Ria is zero (due to computation error, this value is only quite near zero). We can thus use relief intensity average as a depth indicator to adjust our input to the real average depth.

Curie Depths Inversion in Hohhot-Zhang Jiakou Area

Curie depth is a reflection of thermal status in the crust. It is important to geological structure research, mineral resources prospecting and earthquake studies. The northern edge area of north China platform is rich in nonferrous mineral resources and is of great importance in geological study. We introduce here our recent inversion research result using aeromagnetic anomalies in the main mid-west part of the edge, from Hohhot to Zhang Jiakou, 53,200 km^2. Figure 4 (a) shows the distributions of Archean strata, granite rocks and intrusive basalt in this area. Both Archean group and the basalt have high magnetization values. Figure 4(b) is the map of aeromagnetic anomalies.

(1) On the summary of magnetization of different rocks, strata and rough magnetization distribution in vertical in the crust , we obtain the magnetization function of Hohhot-Zhang Jiakou area utilizing genetic algorithm[5].The function is as follows,

$$J(\zeta)=0.3\ \zeta\exp(-0.55\zeta)\quad A/m. \tag{5}$$

(2) We use normalized filter calculation to obtain the anomaly of Curie depth interface after reduction-to-the-pole. Filtering normalization factor is chosen as L=40 km.

(3) Collecting all the known geological information[7,11,13] and seismic reflection and refraction interpretations, and fully utilizing the known depths of upper interface of magnetic crystalline basement, we draw a rough upper interface of magnetic crystalline basement. The average depth of   the basement is 3.2 km to the west of longitude 114°E and 3.5 km to the east.

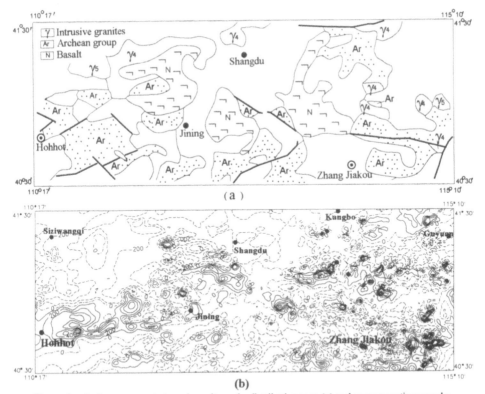

Figure 4. Archean group strata and granite rocks distribution map (a) and aeromagnetic anomaly map (b) of Hohhot - Zhang Jiakou area, north China. Anomaly data unit is nT.

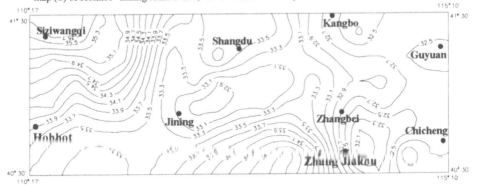

Figure 5. Calculated Curie depth (bottom of magnetic crystalline basement) variation in Hohhot - Zhang Jiakou area, north China. Depth data unit is km.

(4) The average depth of Curie interface is adjusted to 33.0 km. With an accuracy of 5%, we obtain the Cuire depths inversion result, see figure 5. The main upswelling ranges of Curie interface are coincident with main deep fault structures in this area.

Upper Interface Depth of Archean Group in Cai Jiaying Area

Nonferrous ores, such as gold, silver and lead-zinc have much to do with   Archean group (and sometimes Proterozoic group), deep faults and granite rocks (especially buried Yanshan period granite). The Archean group is very thick and has relatively very big magnetization value compared to its overlying strata. This is a quite suitable case to apply aeromagnetic anomalies to reveal buried geological structures and determine upper-interface depth of Archean group.   We have studied 26,112 km^2 around Cai Jiaying mining area. Cai Jiaying mine is nowadays a focus of attention of many geologists and mineralogists.

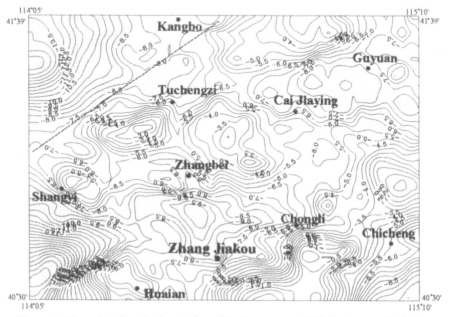

Figure 6. Upper interface depth variation of Archean group in Cai Jiaying area, Heibei, China. Dash line is a interpreted deep fault. Depth data unit is km.

After we filtered the 1/200,000 aeromagnetic anomaly (see figure 4b) data to remove interference of shallow small magnetic bodies and utilized the Curie depth calculated above[12], We completed upper interface depths inversion of Archean group. The result is shown in figure 6. The depth variation shows almost the same up and down steps as that of gravity inversion result using genetic algorithm[3]. We give an interpretation here to the inversion result with consideration of geological structure, igneous rocks and gravity anomalies.

(1) In the south of Cai Jiaying, Tuchengzi and Guyuan, there exist up-swelling areas.
Cai Jiaying mine lies in a depression. The ore bodies discovered so far are lying in a relatively flat area in the middle of the slope of the depression. This depression corresponds to a low gravity anomaly. There must exist a buried Yanshan period intrusive granite between this depression and the ground surface, because density of the granite is relatively small. And, we can see thin stock of this granite in the ground.
There is a similar area to the south of Tuchengzi as well. These two areas are potential prospecting sites.
(2) Cenozoic basalt to the northwest of Zhangbei and Tuchengzi is relatively thin. It is

clearly shown both on aeromagnetic anomaly map and with our filtering result. The doming of Archean group in the south of Tuchengzi is beneath this basalt layer.

(3) There should exist a deep fault along the line from Shangyi to Kangbao. This can be learnt both on aeromagnetic anomaly map and from the inversion depth of Archean group. Characters of magnetic anomalies and depth elevation are totally different between the two sides of the fault.

Acknowledgements

We thank Mr. LIU Guangding, academician of the Chinese Academy of Sciences for his support and directions. Thanks are also due to associated professor LIU Shiyi whose suggestions in geophysical interpretations of Cai Jiaying mining area are very constructive. We are grateful to professor Tan Chengze et al who work in the paleo-magnetism laboratory of China University of Geosciences, lecturer Yao Changli and research fellow Hao Tianyao for their helps in our work.

References

1. Kelochihovskarya, 3.A.,Magnetic nonhomogeneity of the crust, *Journal of Geophysics*(in Russian), **8**(5),3-23(1986).
2. Guan Zhining, An Yulin, *Regional magnetic anomaly quantitative interpretation*, Beijing: Geological Publishing House (1991).
3. Guo Wulin, Petrologic and petromagnetic interpretation to regional magnetic anomalies, *Geophysical and Geochemical Exploration abroad* (in Chinese), **51**(4), 26-32(1987).
4. Parker,R.L., The rapid calculation of potential anomalies, *Geophysics, J.R.Soc.*,**31**,447-445(1973).
5. Zhang Minghua, *New nonlinear data processing methods and applications in Huabei platform area*[PhD. Thesis]. China University of Geosciences,Beijing(1996).
6. Hanson, R.O., Nationwide Curie point depth analysis of Japan, *53th annual meeting, SEG* (1983).
7. Aero-geophysical and remote sensing center, China, *Reports on the surveys and interpretations of gravity and magnetic anomalies in the northern edge of Huabei platform*[research report], Beijing: Aero-geophysical and remote sensing center (1993).
8. Hanson, et al., Magnetic anomalies in Japan and its adjacent regions, *Special issue, Journal of Geomagnetism and Geoelectricity*, **46**(6),(1994)
9. Mu shimin, Shen Ninghua and Sun Yunsheng, *Method of regional geophysical data processing and application*, Changchun: Jinlin Science & Technology Press (1990).
10. Parker, R.L. and Huestis, S.P., The inversion of magnetic anomalies in the presence of topography, *Journal of Geophysical Research*, **79**(11),1587-1592(1974).
11. Wang Guanfu and Zhang Wenzhi, Geological causes of gravity and magnetic anomalies in the northern edge of Huabei platform, *Geophysical and Geochemiacl exploration*(in Chinese), **16**(1), 31-37(1992)
12. Wang Jilun, Two dimensional optimal linear filter design, *Acta Geophysica Sinica*, **20** (2),159-172(1977).
13. Yu Qinfan and Ma Xingyuan, Image process of aeromagnetic anomalies and seismic structure interpretation in Huabei district, *Seismic Geology*(in Chinese), **11**(4), 5-14(1989).

Proc. 30th Int'l Geol. Congr., Vol. 20, pp. 148-157
Liu (Ed.)
© VSP 1997

The New Advancement of Seismic Technique in Coal Industry of China

SHI ZHUOZHOU, TANG JIANYI, FANG ZHENG
China National Administration of Coal Geology, ZhuoZhou, Hebei 072750, China

Abstract

Seismic technique has been used in coal industry of China for 40 years.
Several tens new coalfield and coal production areas were found by this technique in the past. Combining with drilling boreholes, it has finished more than 200 areas with primary, detailed, or precise exploration reports. That leads to the formation of integrated Exploration Technique especially suitable for the geological environment of China.
In recent years, a new seismic technique focused on mining design service has been developed to meet the demands of integrated machinery mining technique from large or extra — large coal mines including Huainan, Huaibei, Pingdingshan, Shenmu, Luan, and others. This new advancement supplies more reliable and more accurate geological results for coal production and coal mine construction.
This paper illustrates the advancement of seismic technique in coal industry of China with practical examples, which includes high resolution seismic, coal production seismic, high precision 3D seismic, lateral coal predicting, VSP, and cavern seismic technique.

Keywords: Coal Seismic, Advancement

INTRODUCTION

China is one of the few nations over the world which take coal as their main energy source. Coal industry in China roughly takes 70% of primary energy production and consuming structure. The coal in China is mainly produced form Permian, Jurassic, and (small portion) Tertiary formations. The relatively complicated geological structure and weak stability of coal deposit cause many problems for coal production. In Eastern and Central China, big machinery mining work faces are frequently affected by geological structures of stopped; Many mining designs are redone or readjusted; Roadways are extended or abandoned.

That causes heavy loss. To use more drilling boreholes in controlling detailed geological structures needs long time and high investment, which not only is non — economic but also is impractical in some cases.

Presently, successful exploration in Eastern and Central China performed with detailed seismic work combining with new technique advancement has achieved good geological results and economic benefits for more than 100 coal mining areas.
This achievement gives these coal mines a chance to reverse the passive situ-

ation related to unknown geological environment during coal mine construction and coal production. So, it gives benefit to the development of coal industry of China.

Seismic technique at present can solve the following geological tasks in coal in dustry:
1. Define faults with throw bigger than 10 meters (2D seismic); Define faults with throw bigger than 5 meters and faulting points for the faults with throw bigger than 3 meters (3D seismic).
2. Define curved structures in main coal seams with amplitude bigger than 10 meters (2D seismic) and 5 meters (3D seismic) with depth error for floor or roof less than 2% (2D seismic) and 1% (3D seismic).
3. Determine and predict coal splitting or combination zone, erosion zone, and cake zone for main coal seams.
4. Determine the position of abandoned road way.
5. Detect cavern.
6. Detect underlying coal crop – out with horizontal position error less than 50 meters.

This paper uses practical examples to illustrate seismic technique advancement in coal industry of China, which includes high resolution seismic, coal production seismic, high precision 3D seismic, coal seam lateral predicting, and cavern detection.

HIGH RESOLUTION SEISMIC TECHNIQUE

High resolution technique in coal mine is a systems engineering, which includes improvement on field work, equipment and data processing. The key factor on increasing resolution is to raise high frequency signals. Certainly, it is not enough to increase frequency only. It must extends the frequency range of seismic signals also and improves signal to noise ratio combining with noise reduction technique. That will raise following problems:
1. How to excite high frequency, wide spectral seismic signals.
2. How to acquire such a signal without significant loss of high frequency seismic components.
3. How to record high frequency reflection information.
4. How to improve signal-noise ratio while preserve high frequency signal.
5. How to compensate the high frequency decaying.
6. How to increase resolution during data processing.

In Eastern and Central China, the solution for these problems are determined by field practice in extra-large coal mines as below:

Exciting
For a suitable signal-noise ratio, high speed explosive is adapted in small charge, usually 0.5—1 kg. Exciting the dynamite in optimum formation that is determined by water – table, formation structure, excited signal high frequency, and relative high signal – noise ratio. The depth of hole is 8—15m or 30m, or using vibroseis in gravel areas with parameters 10—125 Hz, 8—10 sweep.

Acquisition
a. Using 60 Hz or 100 Hz geophones or special geophones, buried in 30 cm or 2 m shallow holes or in 8 m holes, in reducing the decay of high frequency signal caused by surface layer and low speed layer.
b. Sampling rate: 0.5, 1 ms in time; 2.5 m, 5 m, 10 m in space.
c. Using 24 A/D, extra low noise seismic recording system.
d. low cut filter in recorder is 30 Hz, 60 Hz, or 90 Hz.
e. 12, 24, or 48 folds.
f. Multiple geophones for each geophone point.

Data Processing
On the raw records, seismic signals from coal seam reflection have a dominant frequency about 60 Hz. Such a low frequency signal can not reach high resolution. In fact, high resolution problem can not be solved completely in acquisition stage. It is not economic, also. Detailed data processing can increase resolution significantly. Unfortunately, many processing methods reduce resolution, such as stack, although data processing can increase resolution. So, data processing is a very important stage for each area and get heavy attention:
a. Precise static correction, estimated by first break of refraction.
b. Highly accurate moveout correction to reduce distortion and correction error for high frequency components.
c. Noise reduction.
d. Compressing wavelet and reducing its action. Such as: least square statistic deconvolution, Q compensation, wavelet processing, cascade deconvolution, and anti − Q filtering.
e. Borehole constrain inverse: wavelet transformation.
Using these methods can make the dominant frequency of coal reflection reaching up to 100 Hz for 1000 meters less formation with good signal − noise ratio over spectral range 10—200 Hz. Small faults with throw bigger than 5—10 m roughly and coal seam thicker than 0.7 m can be distinguished in time section.

Case History
figure 1 is a Typical High resolution Seismic Section.
figure 2 is a Borehole Constrain Inverse Section.

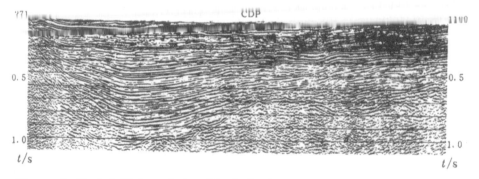

Figure. 1. Typical High resolution Seismic Section.

CDP

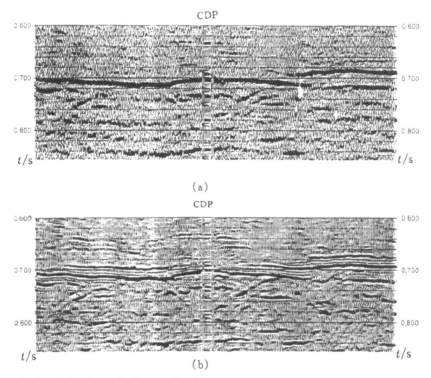

(a)

(b)

Figure. 2. Borehole Constrain Inverse Section.
(a) before inversion; (b) after inversion

3D HIGH RESOLUTION SEISMIC TECHNIQUE

As we have mentioned earlier, high requirement for the details of geological
structures and low mining depth (less than 1000m) give out some special features for 3D seismic technique of coal industry which is rather different with
the 3D seismic technique used in oil and gas industry:

1. Short spread, usually 500—700 m, less than 600 in off-line offset.
2. Dense CDP grid, usually $(5-10m) \times (10-20m)$.

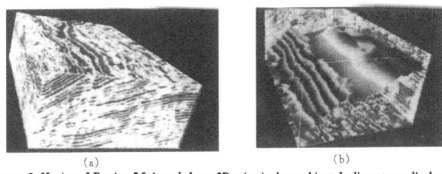

(a)   (b)

Figure. 3 Huainan LB mine 3&4 work faces 3D seismic data cabinet. beding stereo display
(13-1 coal seam reflection)

3. Using high frequency geophones with natural frequency 60 Hz or 100 Hz and buried in depth of 20 – 30 cm.

4. 12 folds CDP data seismic acquisition with 4 line 6 shots or 8 line 3 shots.

5. Calibrating with borehole data 5 – 10 per square kilometers during data

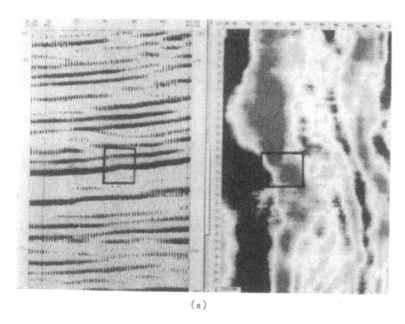

(a)

(b)

Figure. 4 Time slice for Huainan LB coal mine 3D seismic, small fault with throw 5m. (a) (b) . the part in the frame is the small fault interpreted, as the picture showing, the fault point appears very chearly.

processing process.

6. Using velocity data calibrated by borehole to make explanation accurate up to 1% error or less.

7. Dynamic management of seismic results data, that is using mining data to improve seismic explanation results or re – explain the data.

There are 12 areas, each one is 2 km^2 — 10 km^2, having done 3D seismic exploration during past years in Huainan, Huaibei, Jining, Kailuan, Yongxia, Datun, and other coal mines.

The Main Achievements are:
a. Defined faults in mining area which have throw bigger than 5 meters, the faulting points with throw bigger than 3 meters can been seen on seismic section.
b. Depth is accurate up to 1% error or less for floor or roof after actual mining.
c. Determined the main roadway position.

Case History
figure 3 is a Huainan LB mine 3 & 4 work faces 3D seismic data cabinet. beding stereo display(13-1 coal seam reflection)
figure 4 is a Time slice for Huainan LB coal mine 3D seismic, small fault with throw 5m.

COAL SEAM LATERAL PREDICTING SEISMIC TECHNIQUE

Lateral coal seam prediction is a seismic technique to predict lateral coal seam variation with the help from borehole data and well – log data. Usually, the following techniques are used to predict coal lateral variation:
1. Precise calibrating coal seam layer techniques.
2. Geometric description of coal seam floor space features techniques.
3. Thickness lateral prediction of main minable coal seams techniques.
4. Description of coal splitting, coal combining, and erosion zones techniques.
5. Crop – out position predicting technique of main minable coal seams techniques.

The foundation and base for lateral prediction is amplitude variation, phase shifting, frequency variation, and velocity changes.
The reflecting layers of seismic section can be calibrated precisely by synthetic seismic records and vsp data, with this done, coal layers are usually predicted by the following methods:
a. Waveform analysis
b. Characteristic parameters
c. Steady iterative inversion in time domain
d. Seismic trace integration and wave impedance inverse
e. Ratio of trace amplitudes

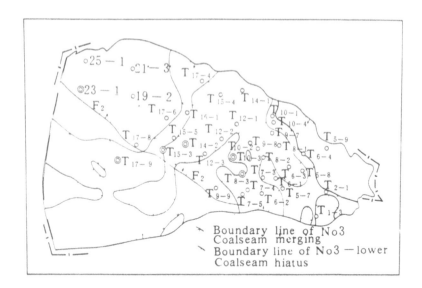

Figure. 5 Using seismic data to explain No.3 coal seam splitting, merging. The results have been confirmed by 38 boreholes at the success rate about 84% in Figure ◎ .drilling and seismic interpreting discord.

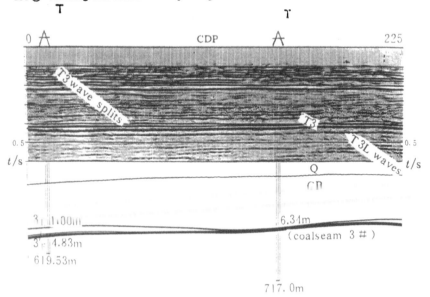

Figure. 6 Typical seismic section showing coal splitting and combining.

Case History
figure. 5 is a Using seismic data to explain No.3 coal seam splitting, merging.. The results have been confirmed by 38 boreholes at the success rate about 84% in Figure ◎ .drilling and seismic interpreting discord.

figure. 6 is a Typical seismic section showing coal splitting and combining.

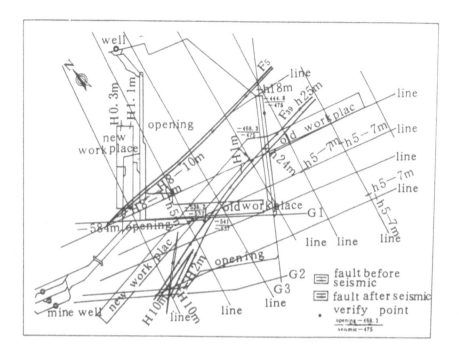

Figure 7. LB coal mine. 3. 4 digging zone, geologic structure camparison, H · h in figure 7 represents fault throw

COAL MINE PRODUCTION SEISMIC TECHNIQUE

Coal production seismic technique is the seismic work done after primary design of coal mine, or after coal mine has been put into production, to solve the geological problems tightly related to laydown work face or pre – work face, or work face. It is developed to meet the demands coming from coal mines in 1990's to serve coal production with the main features as below:

1. Generally adopt 2D or 3D seismic technique.
2. Survey grid is dense. 175×250 m usually for 2D, 125×200 m for difficult areas; 20 – 40m x 40 – 60 m for 3D seismic.
3. main tasks for production seismic are:

a. Define faults with throw bigger than 10 m, fault points with throw bigger than 5m for 2D seismic; Define faults with throw bigger than 5m, faults points with throw bigger than 3m for 3D seismic.

b. Depth error for main coal seam floor is less than 1%.

c. Determine flushing zone for main minable coal seams.

d. Determine cavern range.

Example 1. Huainan LB Mine
This mine is designed to reach 3 million tons output per year and seismic was

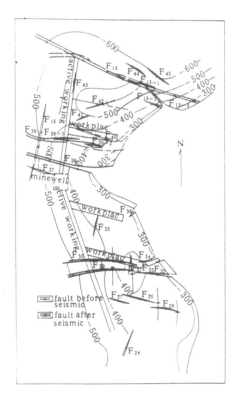

Figure 8. LE coal mine. first digging
zone, geologic structure camprison

done before its finish of construction. On original geologic report,
there is a fault F, so two work
faces were designed around its two
sides. Seismic results show that the
general geological features are in
accordance with the original geological report. However, the faults
position is significantly different as
it shows that both work faces are
cut through by fault F_{39} as showing
in Fig. 7. Hence, two work faces
were adjusted with 3000m forward
space. This coal mine has reached 2
million tons year output after its
production in 1993.

Example 2. LE Mine
This mine is designed to reach 2.4
million tons year output. Original
report shows no faults. Seismic
work finds out 17 faults and indicated out that two work faces of
three original faces are cut through
by faults, referring to Figure. 8.
So, the original design is redone
and eliminated heavy loss. Figure 9
is a typical production zone seismic
section.

CAVERN IN ORDOVICIAN SEISMIC DETECTION

Water in Ordovician cavern is the main obstacle for water prevention and
mining coal seam of Taiyuan Group in Xingtai, Fengfeng, Jiaozuo, Hebi,
Handan, and other mines. Drilling methods take long time and low success
rate in the past, So, cavern seismic detection technique featuring middle or
low frequency, high coverage, and special data processing technique is developed to detect the interior of ordovician limestone, cavern rich zone, the
depth of Ordovician formation, and water leaking features of faults.

CONCLUSION
The paper summarized the new advancement of seismic technique in coal in
dustry of China and its achievements in recent years in coal production. Now,
high resolution 2D seismic, 3D precise seismic, seismic trace inverse, VSP
and so on, especially high precision 3D seismic featuring high resolution,
short work time, are becoming an important tools for coal mine design and
coal production in Eastern China and is expected to become a conventional tool
for coal mines.

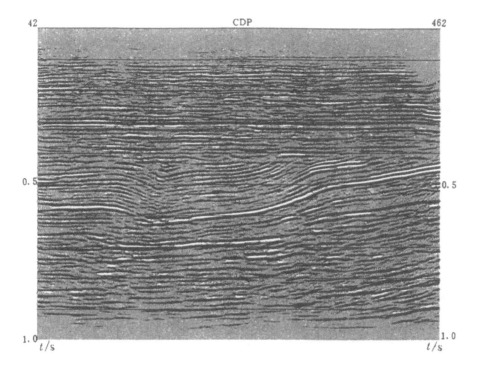

Figure.9 Typical production zone seismic section.

At near future, Eastern and Central China will still be the main energy source of coal, with mining depth increased to 1000m – 1200m. Seismic technique servicing coal industry well try to solve smaller faults with throw less than 3 meters. High resolution and high precision 3D seismic will be developed further to detect coal seam directly. When many data sources and techniques are adapted, or integrated, seismic method will develop towards lateral coal description, coal roof & floor rock feature description, tectonic stress prediction, methane rich zone prediction, faults water leaking features prediction, and play an important role in coal industry.

REFERENCES

1. Tang Jianyi, Coal impedance section, *Coal geology & Exploration*, **3**, 51 – 61(1985)
2. Tang Jianyi, *Seismic section collection of China coal industry*, China coal industry publishing House, **1992**
3. Fang Zheng , Technology of geophysical exploration of coal in China, *Acta Geophysica Sinica* .37 supplement 1, 396 – 403(1994)

Proc. 30th Int'l Geol. Congr., Vol. 20, pp. 158-166
Liu (Ed.)
© VSP 1997

Research of Seismic Methods in Base-metal Ore Exploration

XU MINGCAI[1] HU ZHENYUAN[2] GAO JINGHUA[1] CHAI MINGTAO[1] and
WANG GUANGKE[1]

(1) Institute of Geophysical & Geochemical Exploration, M.G.M.R, Langfang, Hebei, 065000, China
(2) Now a Ph. D. student of China University of Geosciences, Beijing, 100083, China

Abstract

In exploration of base-metal ore deposits, research of seismic method is a very difficult subject. The research was begun with analysis of seismic wave theory and model experiment of the complex wave field and then combined with seismic experiment work of Zouping copper ore deposit in Shandong province. This paper discusses the methods of data acquisition and processing as well as comprehensive interpretations adopted in the research. The results show that mineralization and geological structures can be detected using multi-fold seismic reflection method, while heterogeneity of media related to non-regular copper bearing rocks can be investigated using seismic scattering wave field.

Key words: Base-metal exploration, Seismic method, Multi-fold, Mineralization structure, Scattering wave, Heterogeneity

INTRODUCTION

As the geological prospect for ore extends toward deeper depth, the conventional geophysical method, such as gravity and electromagnetic methods, faces great difficulty due to weak anomaly caused by deep ore body. Hence seismic method based on the elastic wave kinematics and dynamics shows the potential and advantage .

Great achievements have been made and a lot of experience has been accumulated on seismic method for lay-control ore deposits, for example, oil and coal mine. However, seismic method for buried deep hidden metallic ore faced the following problems: complex shape and small scale of ore body, bad continuity of strata, low S/N ratio and severe interference. All those made data processing and geological interpretation even more difficult. Geoscientists of several generations have contributed to study what role seismic method should take in metallic ore survey. Because of its difficulty and lack of financial support, the development was very small in the past decades.

Because of its great detecting depth, high resolution and less ambiguity, seismic method for metallic ore revived. During the 8th 'five year plan', we systematically studied seismic method and technology for metallic ore, and put forward a scheme for determining ore-control structure by reflection or detecting non-regular orebody by scattering wave. The application to Zouping copper ore deposit in Shandong province showed a satisfactory result.

FUNDAMENTAL THEORY OF SCATTERING WAVE

Basic theory

We assume that velocity $c(x)$ of heterogeneous media can be represented by the equation:

$$c^{-2}(x) = c_0^{-2}(x) + f(x) , \tag{1}$$

where x is any point in three-dimensional space, $c_0(x)$ is the reference velocity field of the media, $f(x)$ is scattering potential of the media.

Seismic wavefield approximately satisfies acoustic wave equation under the conditions of the heterogeneous media[1].

$$(\nabla^2 - \frac{1}{c^2(x)} \cdot \frac{\partial^2}{\partial t^2})u(x,s,t) = -\delta(x-s)\delta(t) , \tag{2}$$

where

$$u(x,s,t) = u_0(x,s,t) + u_{sc}(x,s,t) , \tag{3}$$

is the total wavefield, $u_0(x,s,t)$ is the reference wavefield related to the source and reference velocity $c_0^2(x)$, $u_{sc}(x,s,t)$ is the scattering wavefield related to scattering potential, s is the location of a point source.

After FFT with respect to t , considering equation (1) and equation (3), we have acoustic wave equations in frequency domain.

$$(\nabla^2 + \frac{\omega^2}{c_0^2(x)})U_0(x,s,\omega) = -\delta(x-s) , \tag{4}$$

and

$$(\nabla^2 + \frac{\omega'}{c_0^2(x)})U_{sc}(x,s,\omega) = -f(x)\omega^{-1}[U_0(x,s,\omega) + U_{sc}(x,s,\omega)] , \tag{5}$$

When scattering field $U_{sc}(x,s,\omega)$ is far less than reference field $U_0(x,s,\omega)$ in frequency domain, that is, $U_{sc}(x,s,\omega) << U_0(x,s,\omega)$, we can get by Born approximation

$$U_{sc}(s,r,\omega) = \omega^2 \int G(x,s,\omega)G(x,r,\omega)f(x)d^3x , \tag{6}$$

where r is the location of a receiver, $G(x,y,\omega)$ is the Green function of reference media and satisfies:

$$\nabla^2 G(x,y,\omega) + \frac{\omega^2}{c_0^2(x)} G(x,y,\omega) = -\delta(x,y),$$ (7)

When $c_0(x)$ is constant, $G(x,y,\omega)$ can be written as an explicit expression, but when $c_0(x)$ is not constant, the expression is complicated. For convenience, $G(x,y,\omega)$ can be expressed through the first order approximation of geometrical optics as

$$G(x,y,\omega) = A(x,y)\exp[i\omega\tau(x,y)],$$ (8)

where travel time function $\tau(x,y)$ satisfies the eikonal equation:

$$\nabla\tau \cdot \nabla\tau = c_0^{-2}(x),$$

and geometric divergence function $A(x,y)$ satisfies the transport equation:

$$2\nabla\tau \cdot \nabla A + A\nabla^2\tau = 0,$$

Insert equation (8) into equation (6) and let[2]

$$\tau(r,x,s) = \tau(r,x) + \tau(x,s),$$

$$A(r,x,s) = A(r,x) \cdot A(x,s),$$

equation (6) becomes

$$U_{sc}(s,r,\omega) = \omega^2 \int A(r,x,s) exp\left[i\omega\tau(r,x,s)\right]f(x)d^3x,$$ (9)

After inverse Fourier transform, equation (9) becomes

$$u_{sc}(r,s,t) = -\int A(r,x,s)\delta''\left[t - \tau(r,x,s)\right]f(x)d^3x,$$ (10)

From equation (10), one could obtain that the scattering wave field u_{sc}, of which the amplitude is related to $f(x)$, can be formed with energy source shooting if $f(x) \neq 0$. Otherwise, the scattering wave field u_{sc} can't be formed with energy source shooting if $f(x) = 0$. The amplitude of scattering wave has much to do with $f(x)$,

That is to say, the more heterogeneous the underground media, the stronger the scattering wave, and vice versa. Based on the strength of the scattering wave, one could determine whether the scattering wave is produced by ore body or mineralized zone. From equation (10), we know that only the perturbation of the media velocity has a contribution to the production of scattering wave. Therefore, we can study the distribution of the peripheral rock from the wave group characteristic of seismic section.

Modeling experiments
Generally speaking, the geological models of metallic ore exploration are complex. Theoretically all the geological models can be divided into many vertical planks. So it's also equally important to study the seismic response of vertical planks.

For convenience, we designed a 2–D vertical plank model showed in Figure 1 . Figure 2 is the correspondent seismic response. The source is located at the station 48. The diffraction wave at time 100ms is caused by the top boundary of the plank. The amplitude of refraction is strong around the plank. Away from the plank, attenuation of the amplitude becomes significant. Below the refraction wave , there is a low-amplitude but high-frequency type of seismic wave. The range of its appearance horizontally tally with the location of the plank. The low-amplitude and high-frequency seismic wave is a scattering wave produced by the plank. The difference between scattering wave and high frequency interference is that the interference is random, no regularity.

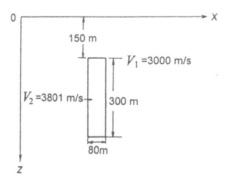

Figure 1 Model of vertical plank

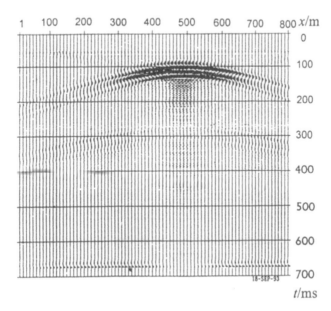

Figure 2 Seismic response corresponding to Figure 1

When the geological models become more complex, physical model experiment is a good tool for complicated media [3]. Figure 3a is a geological model of diffusive block . Figure 3b stands for the result of ultra-acoustic physical model experiment in which zero-offset pattern was used. From Figure 3b we can see, the scattering wave is characterized by high frequency and ordered distribution, similar to the reflection caused by original thin layers.

After analysis of the results of the model experiments, we can infer that the scattering wave caused by the underground heterogeneous media has the characteristic of high frequency and ordered distribution , which are different from those of high frequency interference and reflection. In general, high frequency interference is randomly distributed, but the reflection (or refraction) from underground impedance interfaces shows great coherence to some extents.

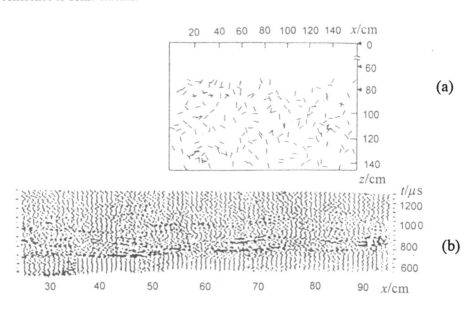

Figure 3 Model of diffusive block (a) and the correspondent result of physical model experiment (b) .

METHOD AND TECHNOLOGY

Zouping copper ore deposit in Shandong province is specially controlled by caldera, in which the pay rock is quartz syenite and diorite, and immediate peripheral rock is volcanic rock. The ore-bearing rock in survey area is located below a 100m deep Quaternary layer.

Before the seismic experiment, we have measured the rock samples to determine physical parameters of the rocks. The result is shown below

Table 1

parameters	rock samples	Quaternary	Quartz syenite diorite	syenite	volcanic rock
velocity (m/s)		1000 –1600	4980 –5220	4750 – 4820	6270 – 6710
density $\times 10^3$ kg/m^3		1.8	2.51– 2.61	2.75 – 2.87	2.67 – 2.84

From the table 1, we can see that the impedance contrast between volcanic rock and quartz syenite diorite and syenite is great. What's more, the impedance contrast between the Quaternary layer and the underlying basement is even greater.

Based on the results measured, combined with requirements of the task of geological survey, and after comprehensive analysis of regional geological, geophysical and geochemical information, we made experiments and did some research work in ore deposit areas.

We carried out multi-fold reflection method and high-fidelity scattering method researches for the ore-control structure and underground heterogeneous media related to ore-body and seismic refraction survey for the analysis of the interval velocity.

After experiments, we chose the operating parameters of multi-fold reflection method as follows: 10m receiver spacing, 100m minimum offset, 12 geophones of 40Hz per trace, dynamite source, ES-2401 seismograph made by Geometric corp., 1024ms record length, 0.5ms sample rate, 35-500Hz band, 24 total traces and 12 multi-fold. The operating parameters of scattering method were the same with the reflection method except for 3m receiver spacing and 150m minimum offset.

The data processing was carried out on a 486 microcomputer and some special software were used. For reflection data, the processing steps included are: demultiplex, editing, amplitude recovery, static correction, two step deconvolution, frequency filtering, decline differential filting, velocity analysis, NMO correction, stack, and so on. For scattering data, the processing steps included are: demultiplex, editing, amplitude recovery, wide band filting, NMO correction, decline differential filting, and scattering wave enhancement. After process, the high quality seismic time sections were got.

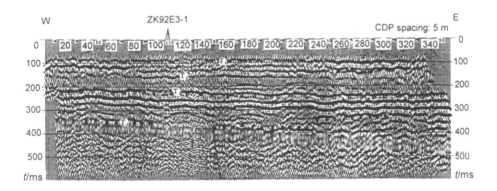

Figure 4 A reflection seismic time section

RESULTS

Figure 4 is a reflection seismic section across a geophysical anomaly at Zouping copper ore target region. Figure 5 is an instantaneous frequency section corresponding to Figure

4 . From Figure 4 , we can see that the strata are smooth and slightly incline to the east. The events T_1 and T_2 are the internal reflection of Quaternary layers. The event T_3 corresponding to the reflection of Quaternary bottom boundary has a characteristic of high amplitude and good continuity. The boundary is about 130–140 m deep. the event T_4 is broken at CDP 235. According to the geological information and the results of gravity and magnetic data, the event T_4 is interpreted as the reflection from an interface between the volcanic rock and median or acid buried rock intruding from caldera. the event T_4 disappears between CDP 95 and CDP 148, which corresponds to the high frequency part at the instantaneous frequency section of Figure 5 . From Figure 4 and Figure 5 , we can detect the location of caldera vent .

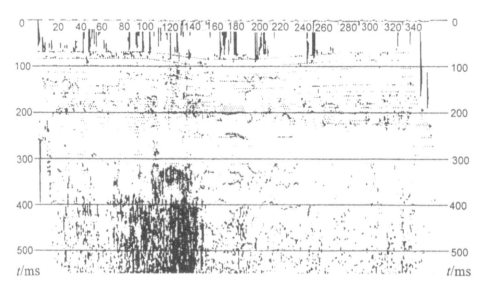

Figure 5 Instantaneous frequency section corresponding to Figure 4

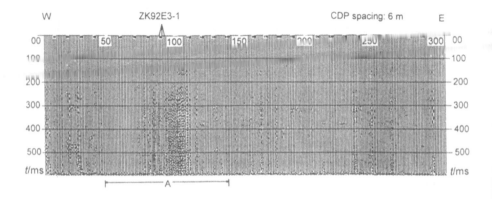

Figure 6 Scattering wave section

Figure 6 is a scattering wave section of Figure 4 . The CDP spacing is 5m in Figure 4 , but 6m in Figure 6. The scattering wave section of Figure 6 was not mixed, such as horizontal stack, so it reflects the true location of heterogeneous media and the degree of heterogeneity.

From the scattering wave section, there is a good coherent and approximately layered high frequency seismic wave, cut the location of the caldera vent of Figure 4 . How to interpret it ? For better study and interpretation, the section of Figure 6 was enlarged to Figure 7. After analyzing the seismic scattering wave section (1.5 m CDP spacing), we have the strong impression that the seismic events inside the caldera vent are very different from that of Figure 4 , outside the caldera vent, randomly distributed low amplitude, but inside the caldera vent, high amplitude of ordered distribution.

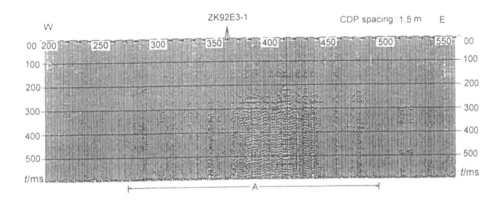

Figure 7 Enlarged section corresponding to A part of Figure 6

From the results of theoretic research, model experiments and regional geological information, we believe that the high frequency seismic wave of Figure 7 is weak scattering wave produced by the heterogeneous media inside the caldera vent. From equation (10) we knew, the more heterogeneous the underground media, that is to say, the greater scattering potential, the stronger the scattering wave. Therefore, based on the strength of scattering wave of Figure 7, we can determine the heterogeneity of underground media.

As we know, the copper deposit is located inside the caldera vent. Generally, the heterogeneity of mineralized zone is weaker than the pay rock. Thus, the strength of scattering waves formed by mineralized zone is smaller than the pay rock. One could differentiate experimentally between mineralized zone and pay rock according to the strength of scattering wave amplitude .

Besides, the location of the caldera vent determined by seismic method tallies well with the results of regional gravity and magnetic anomaly and geochemical anomaly. The range of the caldera vent determined by seismic method is smaller and more accurate than that detected by gravity and magnetic anomaly. The geochemical anomaly and heterogeneous media of seismic exploration have a deviation.

Based on comprehensive analysis of mineralization, geophysical and geochemical anomaly as well as seismic exploration results in the survey area , we could believe that the great heterogeneous media from CDP 370 to 420 may represent the copper ore body. Actually, the depth of Quaternary bottom boundary and mineralized zone revealed by well ZK92E3–1 verified the results of seismic method, but the well deviated from the seismic heterogeneous media. If a new well in the seismic heterogeneous media is drilled , a new ore deposit may be found.

CONCLUSIONS

Through the experimental researches at Zouping copper ore region, the following conclusions can be drawn:

1. Seismic exploration for buried deep ore deposit is applicable if reasonable measurements are made.

2. Seismic reflection method is good at study of ore-control structure, delineation of strata and 3–D geological mapping. Scattering wave method is useful for the study of underground ore related heterogeneous media, furthermore, determination of the location of ore deposit as well.

3. Because of the complex geological conditions, based on the following three ways: theoretical researches, model experiments and many field experiments, seismic methods in base-metal ore exploration can achieve valuable research results.

4. Comprehensive analysis of seismic, regional geophysical and geochemical data would make the location of target area more accurate, make drill more favorable and promote ore-detecting efficiency.

The complex geological structure of metallic ore deposit makes seismic waves complicated. Besides reflection and refraction, diffraction, scattering wave and converted wave are ubiquitous on seismic sections. How to exploit and use that information is an important research subject for better seismic exploration for metallic ore.

REFERENCES

1. Beylin G. Imaging of discontinuities in the inverse scattering problem by inversion of a causal generalized Radon transform, *J. Math. physics*, **26**, 99 – 108 (1985).
2. Miller D., Orislaglio M. and Beylin G., A new slant on seismic imaging migration and integral geometry, *Geophysics*, **52**, 943 – 964 (1987).
3. B.Ja Gelchinsky and N. A Karaev, Heterogeneous seismic models formed by inclusions and investigation of wave field in them *Ann. Geophys*, **36** 519 – 535 (1980).

Proc. 30th Int'l Geol. Congr., Vol. 20, pp. 167-171
Liu (Ed.)
© VSP 1997

Application of Long Window Transient Electromagnetic Method for Hydrocarbon Exploration in Rugged Areas

YAN LIANGJUN, SU ZHULIU AND HU WENBAO

Department of Information Engineering, Jianghan Petroleum Institut, Hubei, 434102,China

Abstract

After overcoming the dificulties of the rugged topography and harsh climate, a field survey of the Long Window Transient Electromagnetic (LOWTEM) method was successfully carried out for the first time in southern China for oil and gas exploration. To make better use of these data, some new methods for data processing and interpretation have been searched. Based on the studies of transient electromagnetic fields over a conducting half space and a conducting thin sheet, two parameters for data interpretation are put forward, one is the all-time apparent resistivity, and the other one is all-time apparent vertical conductance. Pseudo-seismic profiling and 1-D inversion methods are developed also. By using the above methods, the LOWTEM data can trace conducting layers and identify small structures, such as reef buildups quite well. The field trials in rugged areas illustrate that the LOWTEM method is productive and has higher resolution than Magnetotelluric (MT) method, so it has excellent prospects in oil and gas exploration.

Keywords: Transient electromagnetic method, Data interpretation, All-time apparent resistivity

INTRODUCTION

It is very difficult to carry out geophysical survey works in Southern China where the earth surface is covered vastly by limestone with rugged and complicated topography. The conventional seismic exploration methods have been challenged by not only the high velocity surface layer but also the high variation of velocity of the earth media. In an attempt to make breakthrough, several electromagnetic sounding methods have been tested in rugged mountain areas in Hubei, Guizhou and Guangxi Provinces. A so called Long-window Transient Electromagnetic method with long data sampling window, large source power and long offset has been developed specially for hydrocarbon exploration in rugged area. In order to make better use of the LOWTEM field data, great efforts have been put into the field survey design, data processing and interpretation method studies, some encouraging results were obtained.

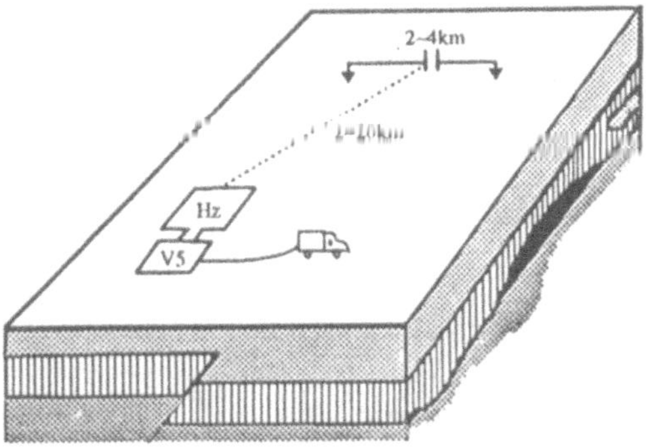

Figure 1. LOWTEM transmitter and receiver

DATA ACQUISITION

The field setup for data acquisition is

illustrated in Figure 1. A grounded wire of approximately 2km in length laid on the earth surface was used as the antenna, a square wave of 30 – 40A current was injected into the ground through the wire, the period of the square wave can be 16s, 32s or 64s. This step current induces eddy currents in the strata and propagates downwards and outwards with increasing time. The PHOENIX-V5 multipurpose receiver was set up at some offset (typically 2 – 20km from center of source) to record the time derivative of the vertical magnetic field response of the earth media immediately after the source current switched off. Numerical model result has shown that for average 100Ωm media, this setup can sustain appropriate signal-to-noise ratio (S/N) in duration of about 4 seconds after current is cut off, and thus an exploration depth of 5 ~ 10km can be reached.

DATA PROCESSING

LOWTEM field data will most likely be distorted by cultural noises, DC shifts (may due to induced polarization effect of some strata), instrument system response and complicated responses of 2-D or 3-D structure. In order to correct these distortions, recursive filtering, DC level out and IP correction scheme have been developed and applied to the induced electromotive force(EMF) curve. Figure 2 shows a selected stack EMF curve after DC level out and IP correction. Obviously, sporadic noises of 50Hz and its harmonics are still retained. After time domain recursive filtering, the periodical noises were almost completely suppressed(see Fig. 3).

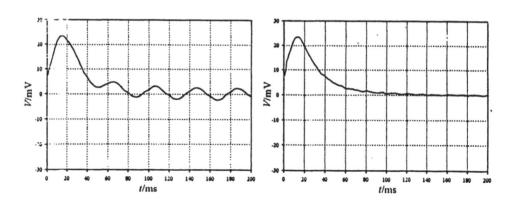

Figure 2. Selected stack EMF data from Guangxi survey area.

Figure 3. Time recursive filtering result of Figure 2.

DATA INTERPRETATION

All-time apparent resistivity
Due to the complicacy of the transient response function, it is difficult to directly define the all-time apparent resistivity by using the EMF response of a homogeneous half-space for data interpretation. Based on the study of characteristics of transient response functions, it has been noticed that the vertical magnetic fileld response is a simple function of the resistivity[1], so that we can use integrated EMF curve to define all time apparent resistivity uniquely.

Figure 4 shows the all time apparent resistivity curve of a three layer earth model compared with the conventional early-time and late-time apparent resistivity curves.

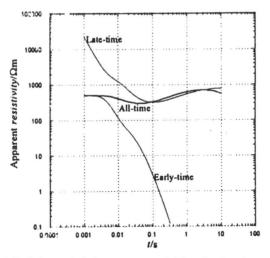

Figure 4. Early,late and all time apparent resistivity of a three layers model:
Thickness(m): h_1=2000, h_2=500; Resistivity(Ωm): ρ_1 =500, ρ_2 =100, ρ_3 =1000

All-time apparent vertical conductance and pseudo-seismic profile

Based on studies of field responses of thin sheet conductor, an approximate formula of all time apparent vertical conductance is derived[1]. It is actually another way of characterization of geoelectrical section, and can be used to qualitatively interpret resistivity structure with better resolution. Since the inflexion points of the apparent vertical conductance curve correspond to interfaces of formations, the time derivative of the all-time apparent vertical conductance can be expressed to imitate the seismic profile and the layer interfaces can be traced through peaks and troughs. Figure 5 is the result of a four layer model, the upper curve is the apparent vertical conductance, and the lower curve is its standardized time derivative. From the time derivatrve curve, formation boundaries can be identified.

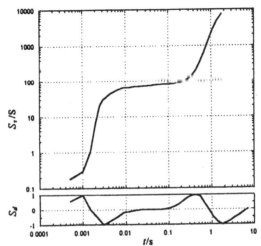

Figure 5. Top curve is apparent vertical conductance and the bottom one is normalized derivative of the top one(S_d is between -1 and +1).

1-D inversion

By means of fast forward modelling algorithm, inversion schemes for layered earth model have been developed for further detail interpretation[2]. Figure 6 shows an example of inverted results for single station field data.

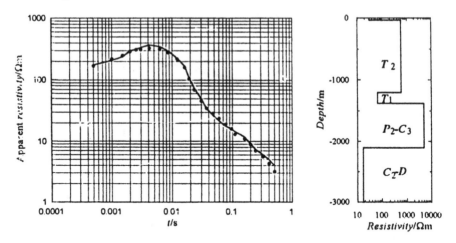

Figure 6. 1-D inverted model with LOWTEM field data(Line: inverted data; Dot: field data).

Generally, the layered model inversion is very computer time consuming since the forward modelling of time domain EM responses requires much more computer time than that for frequency domain modelling. A fast inversion scheme based on all-time responses and continuous model approximation has been proposed, and has been proved to be convenient and reliable compared with the conventional early-time and late-time apparent resistivity approach.

Field examples

In order to map structure anomaly, the interval resistivities of the interested layer were contoured(Fig.7). From this diagram five resistive anomalies can be identified. Figure 8

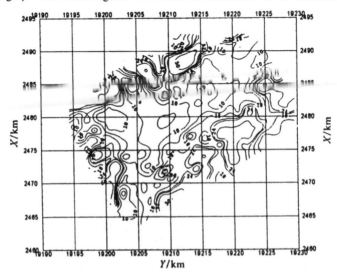

Figure 7. Interval resistivity anomalies in D_2-T_1 formation from LOWTEM data in Guangxi area

shows interpreted seismic anomalies in the survey area. The ring shaped resistivity anomalies in Figure 7 show good coincidence to the eye-ball shaped seismic anomalies in Figure 8, thus the possible interpretation of the five resistive anomalies will be coral reef buildups.

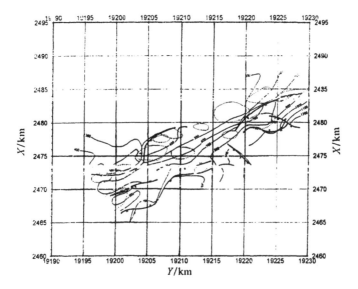

Figure 8. Seismic interpreted geo-structure and anomalies in the same area of Figure 7.

DISCUSSION

The results from our experiments and studies have shown that LOWTEM is probably one of the effective exploration methods in very rugged areas. Our field trials and research works are still in preliminary stage, there are lots of problems require great effort to get satisfactory solutions. It is necessary to carry out 3-D physical modelling or numerical modelling to show how significantly the inhomogeneity around the source grounding points and between the source and receiver affects the observed responses.

In general, the early time S/N of LOWTEM signal is much better than that of MT due to the strong artificial source. For the source capacity, after 4s the late time signal becomes very weak to the limit of instrument sensitivity. This window length is sometime not enough for certain exploration purpose. To improve the S/N at late time, large power supply is required. In Southern China, large power transmitter which can be carried by truck giving a current of 100 amperes is practical, and thus the S/N can be significantly improved compared to the power we have had for our field trials.

REFERENCES

1. Piao Huarong , *The Basic Theory of Electromagnetic Sounding*, Beijing : Ggological Publish House ,1991.
2. Yan Liangjun, Hu wenbao, A random search method in 1-D inversion of LOWTEM data, *Journal of Jianghan Petroliem Institute(in Chinese)*, Vol. 16 Sup., 90 – 92, Dec. 1994.

Proc. 30th Int'l Geol. Congr., Vol. 20, pp. 172-181
Liu (Ed.)
© VSP 1997

Seismic Interferential Ranging Method

LI WENSHAN   SUN WENTAO

Geophysical Prospecting Team, Tuohe, Suzhou City Anhui Province, 234000, China
China National Administration of Coal Geology, Zhuozhou, Hebei, 072750, China

Abstract

The resolution capability of the pulse echo ranging technology is absolutely restricted by the duration of the echo signal. In view of this situation, imitating the principle of Michelson interferometer, author puts forward seismic interferential Ranging method which is independent of the signal waveform and can built with the seismic wave length as measure.

At the beginning in the article, the foundamental principle of the interferential Ranging is expound with simple condition, then it is extended to general conditions and gives out final result. On the basis of this theory the mode tests are successfully completed with computer simulating and some practical data have been processed with good result.

Keywords : Echo Ranging, Interferential Ranging.

INTRODUCTION

The technology of the echo ranging is widely used in the fields of radar, sonar, ultra-sonic detection, and seismic exploration, etc. By using the pulse echo-ranging technology the echo time is read out based on the identified echo signal, then the distance can be converted. Its resolution capability and precision are controlled by the duration and leading-edge gradient of the signal. The conventional technology could meet the requirements in the seismic exploration in the past years when the massive structural outline of the strata was drawn with the marker bed (generally thin-layer composite system) as the rough outline. The requirments for the resolution and precision are higher and reach the internal of the thin layer when new needs for determining the distribution of the thickness and small fault of the coal seam are raised for mining geology by the comprehensive mechanized minining of coal. Since the "thickness" of the coal seam on the time section is generally less than the duration of the echo signal, it is unimaginable to divide further the internal details of the coal seam on the conventional time section with the aid of identifying and reading the echo signal, which is just like to measure the distance shorter than the graduation mark with a common graduated scale. We must disregard the waveform of the signal, the "graduation mark", and change the reference frame.

PRINCIPLE OF THE SEIMIC INTERFERENTIAL RANGING

In 1880's, American physicist Albert Abraham Michelson invented the famous Michelson interferometer, which opened the new way for ranging technology with the wavelength of the light wave as the measure, and improved significantly the ranging precision. Based on the train of Michelson thought, the seismic interferential ranging method independent of the signal waveform can also be built with the seismic wavelength as the measure. In figure 1, S represents the point source on the surface and receiver R is set at the same point; R_i and R_j in the lower part of the figure are a pair of reflectors chosen randomly from the underground reflector sequence. The wavelength (frequency) of the source S is supposed to be controlled.

When the frequency of S is changed to scan R_i and R_j, and the distance difference of echo wave from R_i and R_j is the integral multiple of the wavelength, constructive (or destructive) interference occurs and a brightest (or darkest) line appears on the frequency spectrum , it is on the contrary when the difference of the distance is half wavelength bigger (or smaller) than the wavelength used. The bright or dark line depends on the reflection coeffcents of R_i and R_j with same or reverse sign. Therefore, the frequency spectrum from the scanning result displays bright and dark interferential fringes, and the distance between R_i and R_j can be determined by counting number of time that the fringes appear. In practice, the seismic source contains rich frequency components, the frequency spectrum of the seismogram can be regarded as the record of scanning the strata with different frequency. There are multiple reflections on the real seismic section, among which a pair chosen arbitrarily can be regarded as R_i and R_j mentioned above. Then

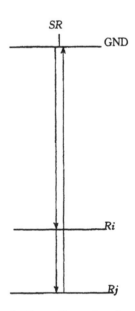

Figure 1. Theory illustration of Intreferntial Ranging

there are C_N^2 groups for combination of reflector pairs for N reflection boundaries. In addition, the spectrum of the real seismic record is complex in combination with the multiple wave. It, in fact, is compounded with the interferential fringes of all "reflection boundary pairs". It is impossible to resolve and read out such complex spectrum with naked eye, finally we must turn to the mathematics for help.

If $F(t)$ and $G(j\omega)$ represent the seismic record trace and its complex spectrum respectively, the Fourier pair of the seismic trace can be represented as following based on the linear superposition and delay principles:

$$F(t) = \sum_{i=0}^{n} k_i f_i(t - \tau_i),$$

$$G(j\omega) = \sum_{i=0}^{n} k_i g_i(j\omega) e^{-j\omega\tau_i},$$

Where $i = 0, 1, \cdots, n$; $j = \sqrt{-1}$; $f_i(t-\tau_i)$ *and* $g_i(j\omega)$ are i-th singal and its complex spectrum; k_i is the strength of the i-th signal ; τ_i is the delay of the i-th signal relative to the reference signal.

This is a common expression, and suitable for all pulse signal including seismic signal. A narrow window near the coal seam can be concerned when we study the coal seam. At this time, it is considered there is only the difference of time delay but no change of waveform for the singal, i. e. it is supposed $f_1(t) = f_2(t) = \cdots f_n(t) = f(t)$, $g_1(j\omega) = g_2(j\omega) = \cdots g_n(j\omega) = g(j\omega)$. If $f(t)$ and $g(j\omega)$ are substituted into the relevant terms of the Fourier pair, following result can be obtained by expansion, elimination and merging under the condition of time invariant.

$$S_0(\omega) = W^2(\omega)[(\sum_{i=1}^{n} k_i \cos\omega\tau_i)^2 + (\sum_{i=1}^{n} k_i \sin\omega\tau_i)^2]$$

$$+ k_0^2 + 2k_0 W(\omega)\sum_{i=1}^{n} k_i \cos(\omega\tau_i - \Delta\varphi(\omega)) \qquad (1)$$

where $S_0(\omega) = G_0^2(\omega)/g_0^2(\omega), W(\omega) = g(\omega)/g_0(\omega), \Delta\varphi(\omega) = \varphi(\omega) - \varphi_0(\omega); G_0(\omega), g(\omega), g_0(\omega), \varphi(\omega), and \varphi_0(\omega)$ are the corresponding amplitude spectrum and phase spetrum respectively.

If $k_0 = 0$ (there is no reference signal), two items on the right of the expression (1) don't exist, then:

$$S_1(\omega) = W^2(\omega)[(\sum_{i=1}^{n} k_i \cos\omega\tau_i)^2 + (\sum_{i=1}^{n} k_i \sin\omega\tau_i)^2] \qquad (2)$$

If
$$S(\omega) = [(S_0(\omega) - S_1(\omega))/k_0 - k_0]/2 \qquad (3)$$

Expressions(1) and (2) are substituted into (3), then

$$S(\omega) = W(\omega)\sum_{i=1}^{n} k_i \cos(\omega\tau_i - \Delta\varphi(\omega)) \qquad (4)$$

$\Delta\varphi(\omega)$ is tried to be omited and let $W(\omega) \approx 1$; if $W(\omega) = 1 + R(\omega)$ (apparently $\int_{-\infty}^{+\infty} R^2(\omega)d\omega \ll 1$), expression (4) is changed into

$$S(\omega) = (1 + R(\omega))\sum_{i=1}^{n} k_i \cos(\omega\tau_i) \qquad (5)$$

Fourier transform is made to the above expression, then based on convolution theorem

$$s(t) = K(t) + r(t) * K(t) \qquad (6)$$

Where $s(t) = FT[S(\omega)]$ is the final output;

$K(t) = FT[\sum_{i=1}^{n} k_i \cos\omega\tau_i] = \begin{cases} k_i, t = \tau_i \\ 0, \text{others} \end{cases}$ is the expected reflection function sequence

$r(t) = FT[R(\omega)]$ can be regarded as the wavelet making up the residual background.

Since $\int_{-\infty}^{+\infty} r^2(t)dt \ll 1$, the last term in the above expression only adds the light residual background on the output channel with $K(t)$ as the main part. Roughly speaking, expression(6) can be regarded as $s(t) \approx K(t)$, i. e. final output $s(t)$ is basically the expected reflection sequence $K(t)$. Thus, an original trace recorded as the waveform is finally inversed into the corresponding reflection time sequence, which breaks away the disturbance of the waveform.

The method is named IR(Interferential Ranging) or PP(Parameter Picking) since it is based on the interferential ranging and characterized with the parameter picking, which is different from the conventional method based on the echo ranging with waveform picking.

MODELING TEST WITH THE COMPUTER

Following are two model tests simulated with the computer. One of the models is a group of reflector sequence composed of eight reflectors. The second model is a thin layer simulating the coal seam.

In Fig. 2, each reflector is located at 80ms, 208ms, 272ms, 304ms, 320ms, 328ms, 332ms and 334ms respectively with the relative interval between the horizon is 128ms, 64ms, 32ms, 16ms, 8ms, 4ms, 2ms, respectively. Each horizon is nearing gradually and tends to the limit of the sampling rate. Fig. 2 (a) shows the supposed geological model, and Fig. 2(b) shows the time section generated from Fig. 2(a) with the sampling interval of 1ms. It is shown from Fig. 2(b) that first and second reflections can be separated without disturbance, the other six are mixed with each other to become a complex wave group and impossible to resolve. Different methods for the improvement of the resolution are applied to make further processing of Fig. 2(b). In order to correlate, the results are listed in Fig. 2 (c)—(g). It is shown from the figures that the different effects are quite evident.

Fig. 3(a) is the supposed (geological) time section(the original model summarizing various geological phenomena, such as pinching out, branching, merging and frequent change in thickness, etc), with thickness change between 0—20ms (matching the actual situation). Fig. 3(b) is the theoretical seismic time section generated from Fig. 3(a), using the wavelet with the apparent period of 20ms(matching the actual situation and thicker than the time thickness of thin layer). It is shown from Fig. 3(b) that it appears as the reflection from continuous, smooth, plicated ridged single reflective "plane" but does not look like coming from a "horizon" with complex internal structure. The compound interference of reflected waves from the top and bottom can't be detected on the whole section except the pinching-out point and merging place where appear two "dark points" because of amplitude anomaly. This shows how difficult to identify a thin layer with time thickness smaller than the apparent period on the conventional waveform section. Fig. 3(c) is the output of Fig. 3(b)after PP processing. The feature of the thin layer in the figure is clear at a glance. The final section can coincide completely with the original model(Fig. 3 (a)) if the "imaginary plane" corresponding to the multiple wave in the lowest part of the section is ignored.

The above two tests prove in theory the accuracy, feasibility, and effectiveness of the method.

EXAMPLES

Following are the results of real data processing.

Firstly it should be explained that the signal of the real data after PP processing is impossible to appear as the ideal Dirac function like that in Fig. 2(c) or Fig. 3(c) but can only be the spike pulse with certain width because of the unavoidable noise and nonideal medium in the real data in addition to the unavoidable artificial factors. The main part of the pulse is the single peak and the position of the peak value is identical to the first arrival of the corresponding wavelet (unnecessary to make the phase correction and it is much easier to identify and pick up than identifing first arrival) on the original section(waveform section). The spike on the real data is not so ideal as that of the pulse without width but much ideal in any case than that of multiple-phase wavelet. First, the single peak(several to more than ten

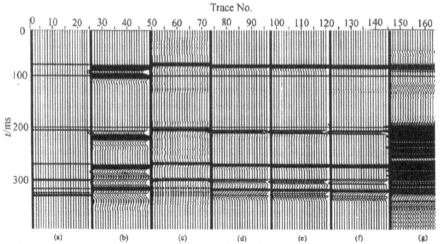

Figure2 The forward time section on a series of closer and closer reflectors and its processing results with different high resolution method

(a) The geological moder(Supposed), (b) The theoretocal time section(generated on (a)), (c) PP processing result of (b), (d) White deconvolution, (e) Pulse deconvolution, (f) Gap deconvolution, (g) Time variable Q compensation.

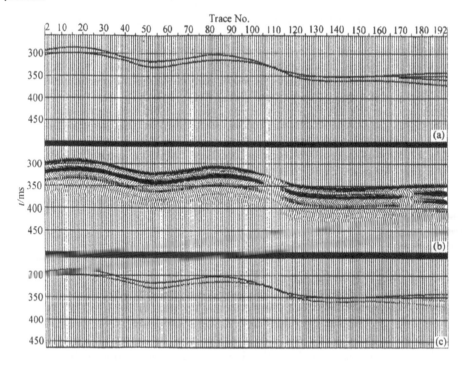

Figure3 The comprehensive model of coal seam structure

(a) Geological model (Supposed), (b) The forward time section(generated on(a)), (c) PP processing result of (b).

samples) is narrower than multiple-phase wavelet (less than 100 samples). Secondly, even if the single peaks interfere with each other because of being too close, the graph superposed by two peaks appears as such simple compound mode as "M" or "trapezoid", which is much easier to identify than the complicated modes superposed with the interference of multiple-phase wavelet. The original peak location may be not found because of the distortion when two single peaks interfere with each other. At this time, half amplitude on both sides of the compound peak can be picked to mark the locations of the top and bottom of the coal seam respectively. But for the depth so determined, the top is about half "half width" shallower than the real depth (depth of the peak) and the bottom is half "half width" deeper. Thus, the thickness between the top and bottom is "half width " more. All of these need to be corrected. Obviously, the precision is poorer if it is quantified based on the position of half amplitude. But it is much better than the waveform section if it is applied qualitatively to draw the form of the coal seam, study the trend of thickness change of coal and detect the faults.

For every segment of the section, the conventional form (waveform) part of the figure is shown on the upper and the parameter section of the same segment after PP processing is shown on the lower part, which is convenient for the comparison. The explanation of every figure is attached to the figure.

The representation of various typical geological phenomenon on the section is shown locally by the above paragraphs (Fig. 4(a)—Fig. 4(f)). A common feature is shown from the figures that the coal seams display details of the structure with specific geological significance on the parameter section, but they are stereotyped with the same appearance of waveform and don't show the "original shape" on the conventional sections. This shows that the internal geological data can 't be revealed to know the structure of the coal seam unless the appearance of the waveform is taken off.

Three complete sections are shown in Fig. 5 – 6 to explain wholly the different representations of the seismic facies in different areas on the two kinds of sections.

A common point can be seen by summarizing these sections that the features of the conventional section in different areas are not prominent, though not stereotyped, since the differences of the internal structure of the coal seam are covered by the similar appearance of the waveform. But it is not so with the parameter section, which has various features and rich details, and reveals various geological information with many appearances.

CONCLUSION

1. The method put forward in the paper is based on the principle of interferential ranging and is a new way for signal processing. It is generally suitable for the processing of various pulse signals including seismic signal.

2. The method has ideal resolution capacity since all components of signal frequency spectrum are fully utilized with effective parameters extracted directly from it and the nuisance of the waveform abandoned. In theory, the resolution capacity is only restricted by the frequency band of the signal and sampling theorem.

3. Basic formula of the method is derived under the condition of time invariant. In

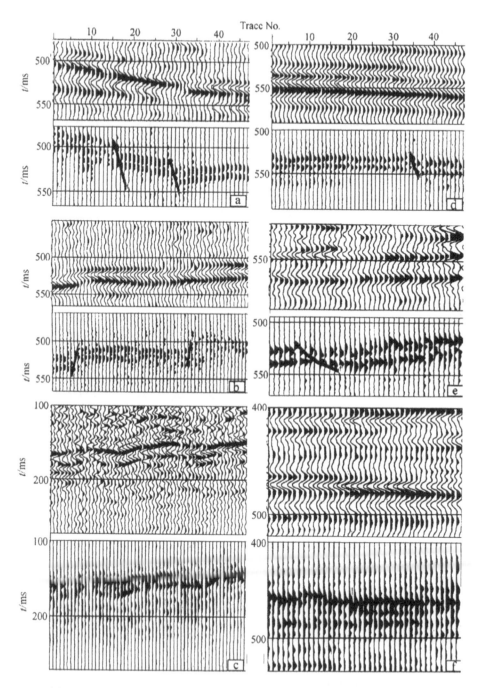

Figure4. (a). The step fault, (b). The small faults, (c). The complex coal lens, (d). The mini − fault in coal seam, (e). The slip sheet structure, (f). A single coal lens.

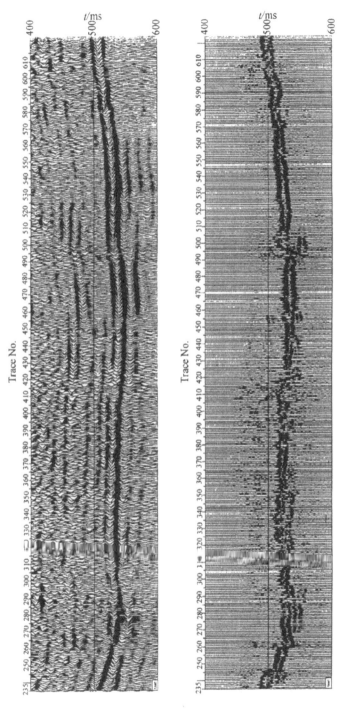

Figure. 5 The distribution of the small faults and the coal seam thickness along the profile

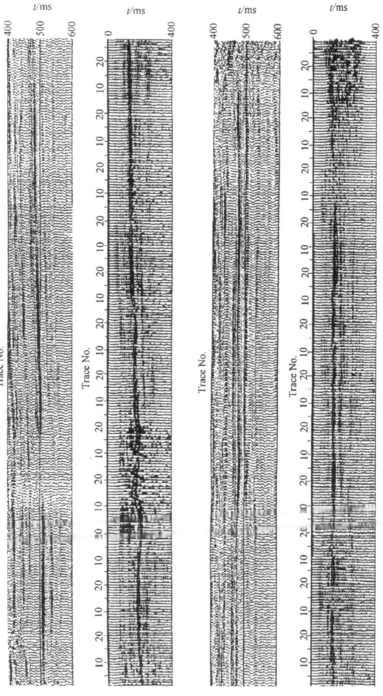

Figure. 6 The seismic facies of the foreset bed

theory, it is only suitable for a narrow window. But time-variant processing can be made for the whole section without real limitations only if the time window is slided with changed parameter and repeated scan along the time axis.

4. The results of the modeling test and real data processing prove the effectiveness of the method and its effects on the display of the internal structure of the thin layer.

5. The method needs to be worked at with great care, and requires higher quality (signal-to -noise ratio, frequency width) of basic data. Real data used in this paper are available data (acquired with conventional method). Thus, the results are not up to the due ideal degree.

6. Only post-stack processing test is made up to now. Better results will be obtained if the original data with high enough signal-to-noise ratio is applied to the pre-stack processing. Firstly, the more original the signal is, the less the loss of the frequency spectrum component is; secondly, parameter stack can avoid the distortion of dynamic correction more than any conventional waveform stack. This is especially important for the shallow layer.

7. The components of the frequency spectrum of the signal shall be guaranteed to be rich enough and be collected in order to tap the latent power of this method. Modern seismograph has very high sampling ratio, but there is still enough space for the improvement of the seismic source and receiver matching the modern equipment and this method.

Acknowledgements

This achievement is impossible without the support from Zhang Yanbing, former Director of CNACG and Lin Xuebing, former Director of Anhui Research Institute of Coal; Thanks are given to Fang Zheng, Jiang Shijun and Mao Bangzhuo, senior engineers of CNACG, for their great help; I'm grateful to Wang Zhenshan, Dean of RICGE, Liu Li and Zhao Xiuli, vice chief engineers of RICGE (Research Institute of Coal Geophysical Exploration) and Gao Yuan, engineer and my assistant for their help in my concrete work; I'd like to also thank Mei Ruwu, senior engineer of office No. 6, No. 3 Institute, Ministry of Neclear Industry, who provides FFT module ; finally, thanks given to Yuan Guotai, leader of Dep of Science and Technology, CNACG, with whose help the paper is published, Here gives my heartfelt gratitude and respects to all fellows.